Fluorine Chemistry Reviews

Volume 6

Fluorine Chemistry Reviews

Volume 6

Edited by *PAUL TARRANT*

Department of Chemistry
University of Florida
Gainesville, Florida

1973

MARCEL DEKKER, Inc.   New York

CONTRIBUTORS TO VOLUME 6

HANS-GEORG HORN, Kehrstuhl fur Anorganische Chemie II, Ruhr-Universitat Bochum, German Federal Republic

MARY T. GLENN, Department of Pathology, University of Florida College of Medicine, Gainesville, Florida

I. McALPINE, Department of Chemistry and Applied Chemistry, Nuclear Science Building, University of Salford, Lancashire, England

PAUL MUSHAK,* Department of Pathology, University of Florida College of Medicine, Gainesville, Florida

JOHN SAVORY, Department of Pathology, University of Florida College of Medicine, Gainesville, Florida

H. SUTCLIFFE, Department of Chemistry and Applied Chemistry, Nuclear Science Building, University of Salford, Lancashire, England

*Present address: Department of Molecular Biophysics and Biochemistry, Yale University, New Haven, Connecticut

iii

CONTENTS

CONTENTS OF OTHER VOLUMES

Fluorine Chemistry Reviews

Reviews

Volume 6

Chapter 1

THE RADIATION CHEMISTRY
OF POLYFLUORINATED ORGANIC COMPOUNDS

H. Sutcliffe and I. McAlpine

Department of Chemistry and Applied Chemistry
Cockroft Building
University of Salford
Lancashire, England

1

I. INTRODUCTION TO RADIATION CHEMISTRY

Radiation chemistry is the study of the chemical effects produced in a system by the absorption of ionizing radiation. These chemical effects may be produced by radiation from radioactive nuclei (α, β, and γ rays), by high-energy charged particles (electrons, protons, deuterons, etc.), and by electromagnetic radiation of short wavelength (x rays of energy greater than 50 eV). This review is concerned only with the effect of γ and x radiation upon polyfluorinated compounds.

Radiation chemistry may be compared to photochemistry, the main differences are found in the energies of the radiation which initiate the reaction and the modes of absorption of energy. The energy of the particles and photons concerned in radiation chemistry is very much higher than the energy of the photons causing photochemical reaction. In photochemistry each photon excites only one molecule; by using monochromatic light of a particular wavelength, a single, well-defined excited state can be produced. In radiation chemistry each photon or particle can (via secondary electrons in the case of photons) ionize or excite a large number of molecules, which are distributed along its track. These higher-energy photons and particles can interact with other molecules giving rise to a variety of possible ions and excited molecules.

The relationship between ionization and chemical action was first put on a firm basis by Lind, who studied the formation of ozone from oxygen [1]. He calculated values of the ion pair or ionic yield (M/N): the ratio of the number of molecules undergoing change (M) to the number of ion pairs formed (N). This was regarded as equivalent to the photochemical quantum yield;

it showed that ionization and chemical action were closely related and proportional to each other. The idea that ions were the sole precursor of chemical effects was opposed by Eyring [2] who drew attention to the fact that the average energy lost in forming an ion pair in a gas (denoted by W) is appreciably greater than the first ionization potential (I) for that gas. It was proposed that the excess energy (W-I) could be used to form electronically excited molecules such as are produced in photochemistry. It is now accepted that both ions and excited molecules are produced and that each can give rise to free radicals which play an important part in the chemical reactions that follow. Further reaction of all these species can, and frequently does, give rise to a complex mixture of products.

Much of the early work on radiation chemistry was with gases, but a great deal of work has since been carried out on condensed systems, especially water and aqueous systems. The work on condensed systems indicated the need for an alternative way of representing the yield. The ion-pair yield (M/N) could not be calculated accurately for condensed systems due to the uncertainty in the value of N for liquids. The alternative was to relate the yield to the energy absorbed, which could be measured directly. The G value was introduced to denote the number of molecules changed for each 100 eV of energy absorbed, and this method is now in general use [3,4].

A. Interaction of γ Radiation with Matter

γ rays tend to lose the greater part of their energy in a single interaction. The result is that some of the incident γ rays are completely absorbed and some continue with their original energy. Hence the number of photons transmitted is decreased and the radiation intensity passing through is reduced. If I is the intensity of radiation transmitted through a thickness x of the absorber then

$$I = I_i e^{-\mu x}$$

where I_i is the intensity of the initial radiation and μ is the total linear
absorption coefficient (cm^{-1}). The total absorption coefficient is the sum
of a number of partial coefficients representing various processes of absorption. These processes are the photoelectric effect, pair production, and the
Compton effect; however, other smaller effects can occur. The relative
importance of each of these processes depends on the energy of the photon
and the atomic number of the stopping material.

1. The Photoelectric Effect

The photoelectric effect is most probable for high atomic number materials
and for low photon energies. It involves the transference of the entire energy
of the photon to a single atomic electron, which is then ejected from the atom
with an energy equal to the difference between the photon energy and the
binding energy of the electron. If the electron comes from an inner shell,
the vacancy created will be filled by another electron from an outer shell
with emission of characteristic x radiation.

2. Pair Production

Pair production involves the complete absorbtion of a photon in the
vicinity of an atomic nucleus to produce an electron and a positron (Fig. 1).
The energy of the photon less the rest energies of the two particles (each
m_0c^2) is divided between the electron and positron. The positron exists
until it has lost its excess energy and then combines with an electron giving
rise to two photons of energy 0.510 MeV which is known as "annihilation
radiation." This effect cannot occur at incident photon energies less than
1.02 MeV, that is, $2m_0c^2$.

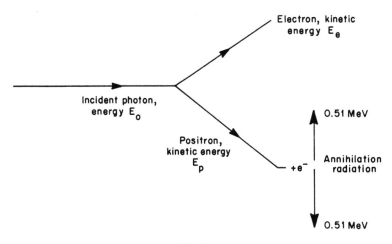

FIG. 1.

3. The Compton Effect

The Compton effect is most predominant for photon energies around 1 MeV for most materials. It is the chief mode of absorption of energy when using a cobalt-60 source, which emits γ rays of two discrete energies, 1.332 and 1.173 MeV, in equal numbers. In this effect a photon interacts with an electron that may be loosely bound or free, so that the electron is accelerated and the photon deflected with reduced energy (Fig. 2). The greater the angle of deflection of the photon, the greater the energy loss of the photon. The

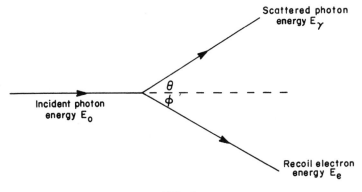

FIG. 2.

energy of the recoil electron is equal to the difference between the energy of
the initial and scattered photon and may have any value from zero to a maxi-
mum, depending upon the angle of deflection of the photon.

The electrons released by the above three processes are called secondary
electrons; they usually have sufficient energy to cause further ionization and
excitation within the medium.

B. Secondary Electron Effects

Electrons lose energy by being slowed down as they pass through a medium.
The chief mode of energy loss is inelastic collisions, that is, through Coulomb
interactions with electrons of the stopping material. Interaction in this way
gives rise to a trail of excited and ionized atoms and molecules.

The rate of energy loss for an electron by inelastic collisions is given by
the Bethe Equation [5] and is known as the stopping power, or specific energy
loss. It is a function of the electron velocity and, therefore, it changes as
the electron is slowed down.

Since electrons of different energies lose energy in matter at different
rates, the concentration of active species in the particle track will vary.
The chemical reactions which subsequently occur depend on this density of
active species and so vary with the particle energy. Expressions which
reflect this changing density of active species are the specific ionization
and linear energy transfer (LET).

The specific ionization is the total number of ion pairs produced in a gas
per unit length of track. The most intense ionization is near the end of the
track where the particle velocity is low.

The LET is the linear rate of energy loss of an ionizing particle traversing
a medium. An approximate average value is calculated by dividing the total
energy of a particle in keV by its path length in microns. For γ rays that
produce secondary electrons with a wide range of energies, the LET will
also vary widely. Track effects are more important in liquids, where the

active species are hindered from moving apart, than in gases, where they can move more freely.

Electrons ejected as a consequence of these secondary electron effects may themselves be sufficiently energetic to produce further ionization and excitation. If the energy of these electrons is less than 100 eV, their range in liquid or solid materials will be short, and any secondary ionization will be situated close to the original ionization giving a small "spur" of excited and ionized species. In water or organic liquids Samuel and Magee [6] have calculated that the spurs occur at intervals of about 10^4 Å and have an initial diameter of about 20 Å. The localized high concentration of energetic species within a spur results in a greater probability of reactions between these species than with the substrate itself. Spur effects are not important in gases because of the low density and ease of diffusion. Electrons which are slowed down to the level of thermal energies will eventually be neutralized by a positive ion directly, or added to a neutral molecule to form a negative ion which itself neutralizes a positive ion.

C. Reactions of Ions and Excited Molecules

The ions and excited molecules, formed by the absorption of ionizing radiation in matter, bring about chemical change by breaking down and/or reacting with the substrate. A knowledge of the breakdown patterns and reactions of these entities is necessary in order that the radiation chemistry process may be understood.

The normal fate of ions is neutralization with an electron or negative ion. Before this can occur electrons must be reduced to thermal energies by interaction with other molecules. The immediate product is a neutral but highly excited molecule.

$$A^+ + e^- \rightarrow A^{**}$$
$$A^+ + A^- \rightarrow A^* + A$$

This molecule usually dissociates to give either molecular products or free radicals, one of which is excited.

$$A^* \ (or \ A^{**}) \rightarrow M^* + N$$
$$A^* \ (or \ A^{**}) \rightarrow R\cdot^* + S$$

Highly excited radicals can also be formed directly by neutralization of a radical ion.

$$R\cdot^+ + e^- \rightarrow R\cdot^{**}$$

These excited radicals sometimes are termed "hot" radicals because they are more reactive than thermal radicals.

If the ion is excited, as it may be if energy greater than the ionization potential of the molecule is absorbed, then the ion can dissociate.

$$(A^+)^* \rightarrow M^+ + N$$

or

$$(A^+)^* \rightarrow R\cdot^+ + S\cdot$$

The charge usually remains with the fragment having the lowest ionization potential, which is generally the larger of the two fragments. Ion-molecule reactions can, of course, occur between ions and the substrate.

$$A^+ + B \rightarrow C^+ + D$$

If the reaction is exothermic, no activation energy is usually required. Alternatively, charge transfer can occur between the ion and a neutral molecule provided the ionization potential of the neutral molecule is equal to or less than the ionization potential of the neutral counterpart of the ion.

Thermal electrons may add to some neutral molecules which are capable of forming negative ions; this is accompanied in some cases by dissociation.

$$A + e^- \rightarrow A^-$$

or

$$A + e^- \rightarrow B^- + C$$

Compounds that show such affinities are the halogens, oxygen, water, and halogenated compounds. For example,

$$C_2H_5I + B^- \rightarrow C_2H_5\cdot + I^- + B$$

Many studies of ions and ionic reactions have been carried out by the use of the mass spectrometer. The mass spectra of organic compounds have also been correlated with the products obtained from them by irradiation in the gas phase [7,8]. These results, however, must be treated with caution because of the vast differences in pressure between gas-phase radiolysis and the ion chamber of a mass spectrometer.

Many excited molecules formed in radiation chemistry are the same as those produced in photolysis but, in addition, ionizing radiation can produce more highly excited states and a greater probability of triplet excited states. Molecules in excited states can lose energy and eventually return to the ground state by radiative conversion (fluorescence) or nonradiative conversion involving lower singlet states or triplet states. Excited molecules can also break down to give most commonly free radicals, or sometimes molecular products.

$$A^* \rightarrow R \cdot + S \cdot$$

or

$$A^* \rightarrow M + N$$

Excited radicals can also be formed if the excitation energy is greater than the dissociation energy of the broken bond.

Thermal free radicals formed by the above processes are themselves reactive intermediates and can react further by radical-radical and radical-molecule reactions.

$$R \cdot + S \cdot \rightarrow RS$$
$$A \cdot + BC \rightarrow AB + C \cdot$$

Radical-radical reactions occur at practically every collision because there is little or no activation energy, whereas radicals may collide a number of times with the substrate before radical-molecule reactions occur. In contrast, excited radicals, because of their higher energy, tend to react on their first collision; radical-molecule reactions are therefore predominant for these species. Free-radical reactions tend to dominate many processes in radiation chemistry and this has been demonstrated by the introduction of free-radical scavengers, electron spin resonance, and pulse radiolysis.

Substances used as radical scavengers are either stable free radicals, e.g., nitric oxide, diphenylpicrylhydrazyl, or compounds that readily give rise to free radicals, e.g., molecular iodine [9]. The use of oxygen, which is a diradical having a triplet ground state, is also of interest. To avoid radiolytic action on the scavengers themselves, they are usually added in low concentration. In liquid-phase radiolysis where spur reactions can occur, low concentrations of scavengers (e.g., 10^{-3} M iodine) will only scavenge radicals that have diffused into the bulk of the medium. If higher concentrations are added, then the spurs may be eroded to some extent and the products obtained are more easily observed. It is generally considered that excited radicals, because they react at the first collision, will not be scavenged by low concentrations of a scavenger. In gas-phase radiolysis Yang has shown that several per cent of nitric oxide is necessary to ensure that sufficient products are formed and also to maintain sufficient scavenger concentration throughout the experiment [10, 11]. However, he has also shown by the use of other additives, that for hydrocarbons, nitric oxide still only scavenges thermal radicals and does not have any effect on nonradical reactions. Recent work suggests that nitric oxide may be effective in scavenging excited fluoro-carbon radicals [12].

The role of the electron in the radiolysis process is of considerable importance and may be studied by introducing into the system a species which has a particularly strong affinity for electrons. Such a process is known as electron scavenging. Compounds which have been used for such studies include nitrous oxide, sulfur hexafluoride, and certain fluorocarbons. This topic is discussed further in Section VII.

Pulse radiolysis involves the irradiation of a compound with an intense pulse of electrons with a view to creating a relatively high concentration of the species formed by the initial absorption of radiation. The technique is linked with a spectroscopic method of identification and study of these species. This technique may be regarded as the radiation chemistry equivalent of flash photolysis in photochemistry.

The rest of this review is devoted to a consideration of the effect of radiation on polyfluorinated organic compounds. It is dealt with, as far as is possible, in order of increasing complexity of the substrate. In many cases results are different for the corresponding hydrocarbons and this may often be attributed to the strength of the C-F bond relative to the C-H bond. In particular, product distribution is often less complex in the case of fluorinated compounds than the corresponding hydrocarbons.

II. COMPOUNDS CONTAINING ONE CARBON ATOM

A. Tetrafluoromethane

Tetrafluoromethane is the simplest compound containing carbon-fluorine bonds only, and its radiation chemistry has been extensively studied [13-17]. The results of this work are summarized in Table 1.

In the work of Fajer et al., considerable care was taken to remove oxygen from the reaction system. The tetrafluoromethane was brought into contact with a Na/K alloy, effectively reducing the oxygen content to 0.03%. The surface of the Monel reaction vessel was prefluorinated, a treatment which should remove adsorbed oxygen and leave a fluorinated surface. After this treatment $G(C_2F_6) \sim 1 \times 10^{-2}$. When the oxygen concentration was increased the hexafluoroethane yield decreased and hexafluorodimethyl ether was the predominant product. The following mechanism was proposed for the radiolysis of perfluoromethane alone:

$$CF_4 \longrightarrow CF_3\cdot + F\cdot \tag{1}$$
$$F\cdot + F\cdot + M \longrightarrow F_2 + M \tag{2}$$
$$CF_3\cdot + F_2 \longrightarrow CF_4 + F\cdot \tag{3}$$
$$CF_3\cdot + F\cdot \longrightarrow CF_4 \tag{4}$$
$$CF_3\cdot + CF_3\cdot \longrightarrow C_2F_6 \tag{5}$$

No mechanism was suggested for the formation of hexafluorodimethyl ether, and elemental fluorine was not detected amongst the products. Reactions (3) and (4) were thought to predominate over (5), thus accounting for the low value of $G(C_2F_6)$. The formation of hexafluoroethane by reaction (5) was substantiated by the reduction of $G(C_2F_6)$ in the presence of oxygen. The low value of $G(C_2F_6)$ is consistent with the ESR data on irradiated tetrafluoromethane. At 77°K in a xenon [18] matrix or in the liquid phase [19], the radiolysis of tetrafluoromethane gives no ESR signal due to trifluoromethyl radicals. The trifluoromethyl radical was detected by ESR in the presence of hydrocarbons, which act as scavengers for fluorine atoms

$$RH + F\cdot \rightarrow HF + R\cdot \qquad (6)$$

Thus, reactions (3) and (4) are reduced.

Askew, Reed, and Mailen [14] irradiated tetrafluoromethane in aluminum vessels with γ rays and found $G(C_2F_6) = 0.26$ and $G(C_3F_8) = 0.023$. The oxygenated products, carbon dioxide, perfluorodimethyl ether, and perfluoro-methylethyl ether, were also found. Calculation of this combined oxygen showed it to be equivalent to 0.3% initial oxygen which was considered to be adsorbed on the surface or present as aluminum oxide. Despite this higher oxygen content, the $G(C_2F_6)$ value is higher than Fajer's. The explanation advanced by Mailen et al. is that reaction (5) can occur in the gas phase and only radicals reaching the walls can encounter oxygen. In Fajer's work the oxygen is present in the gas phase and can scavenge trifluoromethyl radicals.

Clearly the nature of the vessel surface is of considerable importance in work of this type. In glass vessels Feng et al. [16] found that the irradiation of tetrafluoromethane gave $G(C_2F_6) = 2.0$. This is a much higher value than was found by previous workers and may be attributed to attack of fluorine on the glass walls, reducing reactions (3) and (4), and increasing reaction (5). The $G(C_2F_6)$ value is markedly reduced in the presence of chlorine and bromine concurrent with the formation of bromotrifluoromethane, $G(CF_3Br) =$ 3–5, and chlorotrifluoromethane, $G(CF_3Cl) = 0.35$. Feng et al. suggest that the nonscavenged hexafluoroethane arises by the reaction of $CF_2\cdot$ or CF_2^+ with a molecule of tetrafluoromethane. The formation of difluorocarbene does seem unlikely since tetrafluoroethylene or products arising therefrom have not been observed by any workers.

The importance of heterogeneous reactions in tetrafluoromethane radiolysis is highlighted by a study in the presence of carbon using reactor

TABLE 1

Product Yields from the Radiolysis of Tetrafluoromethane

Reactants	G values for								Comments	Ref.
	$-CF_4$	C_2F_6	C_3F_8	CF_3OCF_3	$CF_3OC_2F_5$	CO_2	CF_3Br	CF_3Cl		
$CF_4 + 0.03\% O_2$	--	0.01							CF_4 carefully purified. Prefluorinated Monel vessels. CF_3OF suspected	[13]
$CF_4 + 1\% O_2$		0.001		0.05						
$CF_4 + 0.3\% O_2$	1.14	0.26	0.023	0.27	0.016	0.3			Aluminium vessels. Oxygen assumed to be adsorbed on Al or as oxide	[15]
CF_4		2.0							Glass vessels	
$CF_4 + Br_2$		~1.0					3 − 5			[16]
$CF_4 + Cl_2$		0.0						0.35		

radiation [13]. After four weeks of irradiation in the presence of carbon, only 37% of the tetrafluoromethane was recovered unchanged, whereas in the absence of carbon, 98-99% of the tetrafluoromethane was recovered unchanged. The fluorocarbons C_2F_6, C_3F_8, C_4F_{10}, C_5F_{12}, C_6F_{14}, and C_7F_{16} were also isolated from the reaction with carbon. Similar work in the presence of aluminium powder shows a 93% recovery of tetrafluoromethane together with the formation of small amounts of higher fluorocarbons. Therefore, an increased surface area alone is not sufficient to destabilize tetrafluoromethane and build up appreciable amounts of higher fluorocarbons.

B.   Halofluoromethanes*

All three trifluorohalomethanes have been studied but only trifluoroiodo-methane has been studied in both the gas and liquid phase.

The radiolysis of gaseous bromotrifluoromethane in glass over the dose range $1.42 \times 10^{20} - 61.81 \times 10^{21}$ eV/g gave tetrafluoromethane, $G(CF_4) = 1.44 - 1.53$; dibromodifluoromethane, $G(CF_2Br_2) = 1.25 - 1.78$; hexafluoroethane; and bromine. No quantitative data is available for the last two compounds [14]. The yield of tetrafluoromethane and dibromodifluoromethane both show an increase with temperature at constant dose. Feng uses this data to determine an activation energy of 0.39 kcal/mole for the formation of tetrafluoromethane. In the presence of bromine the yield of tetrafluoromethane is unchanged but the yield of dibromodifluoromethane is increased slightly.

It is suggested that tetrafluoromethane is formed by an ion-molecule reaction,

$$CF_3^+ + CF_3Br \rightarrow CF_4 + CF_2Br^+ \tag{7}$$

but that dibromodifluoromethane may arise by either ion or radical abstraction reactions. Since evidence from photochemical reactions indicates that fluorine abstration by a trifluoromethyl radical does not occur [19,20], such an

*The term halogen refers to chlorine, bromine, or iodine.

abstraction reaction is dismissed in this radiolytic work. Further, since added bromine does not affect the yield of tetrafluoromethane, this is also cited as evidence against a radical abstraction reaction. The ion-molecule reaction is thought to proceed via initial polarization of a CF_3Br molecule by the electric field of a CF_3^+ ion followed by the formation of a partial bond between the CF_3^+ ion and the polarized CF_3Br molecule

$$CF_3^+ \ldots F^{\delta -} \ldots \overset{\delta +}{CF_2Br}$$

Finally, there is simultaneous bond breakage and bond formation to give tetrafluoromethane. The low value for the activation energy of formation of tetrafluoromethane is consistent with this interpretation.

In contrast to the formation of tetrafluoromethane, dibromodifluoromethane may be produced by bromine abstraction by either the ion CF_2Br^+ or the radical $CF_2Br\cdot$. In addition, interaction of these intermediates with bromine atoms or bromine molecules would also result in the formation of dibromodifluoromethane, for example,

$$CF_2Br^+ + CF_3Br \rightarrow CF_2Br_2 + CF_3^+ \tag{8}$$
$$CF_2Br^+ + e + M \rightarrow CF_2Br\cdot + M \tag{9}$$
$$CF_2Br\cdot + CF_3Br \rightarrow CF_2Br_2 + CF_3\cdot \tag{10}$$
$$CF_2Br\cdot + Br\cdot + M \rightarrow CF_2Br_2 + M \tag{11}$$

The radiolysis of chlorotrifluoromethane has also been carried out in glass vessels, and a stringent attempt was made to eliminate foreign matter from the walls [15]. This was done by evacuating the vessel, allowing a small amount of chlorotrifluoromethane to diffuse in, and subjecting the vessel to a discharge from a Tesla coil. This procedure usually, but not always, eliminated hydrogeneous material from the glass walls, which was indicated by the appearance or non-appearance of fluoroform among subsequent irradiation products of chlorotrifluoromethane.

The radiolysis of gaseous chlorotrifluoromethane over the dose range 1.3×10^{20} - 1.9×10^{21} eV/g gave tetrafluoromethane, $G(CF_4) = 2.0$; dichlorodifluoromethane, $G(CF_2Cl_2) = 2.3 - 25.4$; chlorine; and hexafluoroethane. The $G(CF_2Cl_2)$ values decrease with increasing dose, suggesting that the radiation stability of dichlorodifluoromethane is lower than that of chlorotrifluoromethane. The radiolysis of chlorotrifluoromethane has also been carried out in the presence of bromine with a view to (a) scavenging radical intermediates, and (b) eliminating possible interference by the last traces of hydrogeneous material from the vessel walls. It was shown that $G(CF_4)$ remains essentially constant in the presence of bromine, except at high $Br_2:CF_3Cl$ ratios. Two bromine-containing products are formed, CF_3Br and CF_2ClBr. The yields of these two products increase with increasing $Br_2:CF_3Cl$ ratio at the expense of the CF_2Cl_2 yield. The reaction in the presence of bromine is represented by

$$CF_3^+ + Br_2 \longrightarrow CF_3Br + Br^+ \tag{12}$$

$$CF_2Cl^+ + Br_2 \longrightarrow CF_2ClBr + Br^+ \tag{13}$$

$$CF_3\cdot + Br_2 \longrightarrow CF_3Br + Br\cdot \tag{14}$$

$$CF_2Cl\cdot + Br_2 \longrightarrow CF_2ClBr + Br\cdot \tag{15}$$

Although, if CF_3^+ reacted with bromine to give CF_3Br, a reduction in the yield of CF_4 would be expected and this is not observed.

The radiolysis of gaseous trifluoroiodomethane like that of chloro- and bromotrifluoromethane has been carried out in glass vessels [21]. The surface of the vessel was conditioned by heating in vacuo followed by repeated irradiation of samples of trifluoroiodomethane until consistent results were obtained. After conditioning, a series of experiments was completed without allowing air into the reaction vessel. When the glass surface is not conditioned, the products are tetrafluoromethane, hexafluoroethane, difluorodiiodomethane, fluoroform, iodine, carbonyl fluoride, and silicon tetrafluoride. In conditioned vessels the products (and 100 eV yields) are tetrafluoromethane (1.08), iodine (0.13), and difluorodiiodomethane (0.82). It is considered that during the conditioning process trifluoromethyl radicals attack the glass surface, removing oxygen as carbonyl fluoride and leaving fluorine atoms bonded directly to silicon at the surface of the reaction vessel.

In the presence of the radical scavengers oxygen and nitric oxide, the yield of tetrafluoromethane is markedly reduced (see Table 2). In view of this it is considered that the main fate of the ions initially formed is to form trifluoromethyl radicals. Radicals so produced are energetic or excited

TABLE 2

Product Yields from the Radiolysis of Trifluoroiodomethane

Reactants	I_2	CF_4	C_2F_6	CF_2I_2	Products N_2	CF_3NO	CF_3NO_2	NO_2	$(-NO)$	COF_2	CF_3OCF_3
Gaseous CF_3I	0.13	1.08	0.0	0.82	—	—	—	—	—		
Gaseous CF_3I + 5% O_2	9.6	0.04	0.08		—	—	—	—	—	8.0	0.51
Gaseous CF_3I + 5% O_2	5.9	0.28	0.26		—	—	—	—	—	10.5	0.51
Gaseous CF_3I + 5% O_2	10.2	0.59	0.07		—	—	—	—	—	49.5	1.39
Gaseous CF_3I + 50% O_2	7.5		0.06	0.08	—					20.9	0.0
Gaseous CF_3I + 6% NO	12.1	0.03	0.03		—	27.7		1.3	83		
Gaseous CF_3I + 50% NO	10.1	0.2	0.02		150	0.32	3.3	251	571		
Gaseous CF_3I + 80% NO	12.3				478				1092		
Liquid CF_3I	1.36	0.37	1.03	a							
Liquid CF_3I + 0.1% O_2	4.27	0.06	5.72							Trace	White solid formed
Liquid CF_3I + 2% NO	4.47	0.18	1.86		6.93	0.36	Trace			Trace	
Liquid CF_3I + 4% NO	4.53	0.20	0.97		17.87	1.24	Trace	26.74	71	Trace	

a A small amount of CF_2I_2 is formed at high dosage.

radicals and have considerably more energy than a radical produced photo-
chemically. These excited trifluoromethyl radicals can abstract an atom of
fluorine from a molecule of trifluoroiodomethane to form tetrafluoromethane
and a difluoroiodomethyl radical

$$CF_3I \quad \rightsquigarrow \quad CF_3^* \cdot \quad + \quad I \tag{16}$$

$$CF_3^* \cdot \quad + \quad CF_3I \quad \rightarrow \quad CF_4 \quad + \quad \cdot CF_2I \tag{17}$$

Any tetrafluoromethane formed in the presence of radical scavengers is
attributed to an ionic abstraction analogous to the process described above for
bromotrifluoromethane radiolysis. The difluoroiodomethyl radical formed
in Eq. (17) reacts with either an iodine atom, an iodine molecule, or a
molecule of trifluoroiodomethane to yield difluorodiiodomethane

$$M + \cdot CF_2I + I \cdot \quad \rightarrow \quad CF_2I_2 + M \tag{18}$$

$$\cdot CF_2I + CF_3I \quad \rightarrow \quad CF_2I_2 + CF_3 \cdot \tag{19}$$

$$M + CF_3 \cdot + I \cdot \quad \rightarrow \quad CF_3I + M \tag{20}$$

Consideration of the G values above shows that $G(CF_4) - G(CF_2I_2) =$
0.26 and since $G(I_2) = 0.13$ then $G(I) = 0.26$. Thus for every molecule of
tetrafluoromethane produced in excess of difluorodiiodomethane, an atom of
iodine is also found. This is interpreted in terms of a surface reaction; a
trifluoromethyl radical produced at the surface abstracts an atom of fluorine
from the conditioned surface to give tetrafluoromethane and iodine. The
ratio of carbon to fluorine of 1:3.2 in the products also favors the idea of
fluorine abstraction from the surface.

The radiolysis of trifluoroiodomethane in the presence of oxygen yields
iodine, tetrafluoromethane, hexafluoroethane, perfluorodimethyl ether,
carbonyl fluoride, an unidentified colorless liquid, and a white solid. The
measured G values show considerable scatter (see Table 2) but tentative
suggestions have been made regarding the mechanism of the reaction in the
presence of oxygen. Since the tetrafluoromethane yield is markedly reduced,
the reaction

$$CF_3 \cdot + O_2 \quad \rightarrow \quad CF_3 \cdot O \cdot O \cdot \tag{21}$$

and subsequent reactions of the $CF_3 \cdot O \cdot O \cdot$ radical must play a key role in this reaction. Some possible reactions are

$$2CF_3 \cdot O \cdot O \cdot \rightarrow \{CF_3 \cdot O \cdot O \cdot O \cdot O \cdot CF_3\} \rightarrow 2CF_3 \cdot O \cdot + O_2 \qquad (22)$$

$$\rightarrow 2CF_3 \cdot + 2O_2 \qquad (23)$$

$$CF_3 \cdot + CF_3 \cdot \rightarrow C_2F_6 \qquad (24)$$

$$CF_3 \cdot + CF_3 \cdot O \cdot \rightarrow CF_3 \cdot O \cdot CF_3 \qquad (25)$$

The radiolysis of trifluoroiodomethane in the presence of nitric oxide yields nitrogen, nitrogen dioxide, trifluoronitrosomethane, trifluoronitromethane, tetrafluoromethane, hexafluoroethane, and iodine [22]. The G values are given in Table 2. The reduced yield of tetrafluoromethane suggests that the initial scavenging reaction is

$$CF_3^{*} \cdot + NO \rightarrow CF_3 \cdot NO \qquad (26)$$

The yield of trifluoronitrosomethane depends upon the relative scavenging abilities of nitric oxide and iodine, and on the reactivity and environment of trifluoronitrosomethane. In view of the known tendency of trifluoronitrosomethane to react with nitric oxide, it is expected and observed that, with 50% M nitric oxide, $G(CF_3NO)$ is low. By contrast, with 5% M nitric oxide, $G(CF_3NO) = 27.7$. The overall scavenging reaction is thought to proceed in two stages. The following chain reaction proceeds when large amounts of nitric oxide are present.

$$CF_3^{*} \cdot + NO \rightarrow CF_3 \cdot NO \qquad (27)$$

$$CF_3 \cdot NO + 2NO \rightarrow CF_3 \cdot N(NO) \cdot ONO \rightarrow CF_3 \overset{+}{N}_2 \cdot \overset{-}{N}O_3 \qquad (28)$$

$$CF_3N_2NO_3 \rightarrow CF_3 \cdot + N_2 + \cdot NO_3 \qquad (29)$$

$$\cdot NO_3 + NO \rightarrow 2NO_2 \qquad (30)$$

$$CF_3 \cdot + NO \rightarrow CF_3NO \qquad (31)$$

When the nitric oxide is depleted and the nitrogen dioxide concentration becomes appreciable, the chain termination reaction

$$CF_3 \cdot + NO_2 \rightarrow CF_3 \cdot NO_2 \qquad (32)$$

begins to compete with, and may eventually take precedence over, the scavenging reaction (27). This is supported by the yield of trifluoronitromethane,
$G(CF_3NO_2) = 3.3$.

The ratio $G(-NO):G(NO_2):G(N_2) = 3.8:1.7:1.0$ is close to the ratio of
4:2:1 expected for the disproportion of nitric oxide,

$$4NO \rightarrow N_2 + 2NO_2$$

and suggests that the main reaction is the catalytic disproportionation of
nitric oxide.

The occurrence of this chain reaction and the disproportionation of nitric
oxide casts doubt on the usefulness of nitric oxide as a radical scavenger in
radiolysis work. Possibly a more useful radical scavenger would be trifluoronitrosomethane, known to be susceptible to radical attack at the nitrogen
atom. Such radical attack would be expected to give rise to either a nitroxide
radical, $CF_3(R)NO$, a substituted hydroxylamine, $CF_3(R)NOR$, or products
arising from these species, for example, CF_3R. Other monomeric nitro-
soalkanes may also be of use in this respect.

The radiolysis of trifluoroiodomethane in the liquid phase contrasts with
that in the gas phase in that hexafluoroethane, $G(C_2F_6) = 1.03$, is a major
product [23]. Tetrafluoromethane, $G(CF_4) = 0.37$, and iodine, $G(I_2) = 1.36$,
are also formed. These results are readily explained in terms of radical-
radical reactions within spurs and a greater chance of deactivation of excited
radicals in the liquid phase than in the gas phase.

The formation of tetrafluoromethane, as in the gas phase, is attributed
to fluorine abstraction by excited trifluoromethyl radicals. The lower yield
is due to deactivation of some of the excited trifluoromethyl radicals. The
thermal trifluoromethyl radicals so formed may be scavenged by iodine or
dimerise to give hexafluoroethane:

$$CF_3\cdot \; + \; I_2 \qquad\qquad \rightarrow \; CF_3I \; +\cdot I\cdot \qquad\qquad (33)$$

$$CF_3\cdot \; + \; I\cdot \; + \; M \qquad \rightarrow \; CF_3I \; + \; M \qquad\qquad (34)$$

$$CF_3\cdot \; + \; CF_3\cdot \; + \; M \quad \rightarrow \; C_2F_6 \; + \; M \qquad\qquad (35)$$

The latter reaction will be favored by the high concentration of active species
expected in a spur.

The presence of oxygen or nitric oxide in the liquid-phase radiolysis of trifluoroiodomethane markedly reduces the yield of tetrafluoromethane and increases the yield of iodine and hexafluoroethane. This observation supports the hypothesis of radical abstraction of fluorine for the formation of tetra-fluoromethane but also implies that nitric oxide is scavenging excited radicals. The increased yield of heaxfluoroethane in the presence of oxygen is surprising, although a similar phenomenon was observed in the gas phase. This increased yield is attributed to the decomposition of an intermediate peroxy compound or radical. The proposed reaction sequence is

$$CF_3^{\cdot *} + O_2 \quad\longrightarrow\quad CF_3 \cdot O \cdot O \cdot \tag{36}$$

$$CF_3 \cdot O \cdot O \cdot + CF_3 \cdot \quad\longrightarrow\quad CF_3 \cdot O \cdot O \cdot CF_3 \tag{37}$$

$$CF_3 \cdot O \cdot O \cdot + CF_3 \cdot^{*} \quad\longrightarrow\quad CF_3 \cdot O \cdot O \cdot CF_3 \tag{38}$$

$$CF_3 \cdot O \cdot O \cdot CF_3 \quad\longrightarrow\quad C_2F_6 + O_2 \tag{39}$$

$$2CF_3 \cdot O \cdot O \cdot \quad\longrightarrow\quad C_2F_6 + 2O_2 \tag{40}$$

Thus when a molecule of oxygen enters a spur, the radical is scavenged, followed by the regeneration of oxygen, which is then available for further scavenging. Scavenging, followed by regeneration of oxygen, explains why such a low concentration of oxygen is effective in almost eliminating tetra-fluoromethane as a product.

Two other reactions involving oxygenated species are feasible, and also result in the formation of hexafluoroethane and the regeneration of oxygen:

$$CF_3 \cdot O \cdot O \cdot + CF_3 \cdot^{*} \quad\longrightarrow\quad 2\, CF_3 \cdot O \cdot \tag{41}$$

$$CF_3 \cdot O \cdot + CF_3 \cdot O \cdot \quad\longrightarrow\quad C_2F_6 + O_2 \tag{42}$$

The products obtained from the radiolysis of liquid trifluoroiodomethane in the presence of nitric oxide are similar to those obtained in the gas phase. A similar chain mechanism explains their formation. Reference to Table 2 shows that for the two reactions in the liquid phase, the difference in the $G(CF_3NO)$ values is almost the same as the difference in $G(C_2F_6)$ values. This indicates that a molecule of trifluoronitrosomethane reacts to give a molecule of hexafluoroethane. This is thought to occur by attack of an excited trifluoromethyl radical on the nitrogen of trifluoronitrosomethane,

followed by decomposition of the nitroxide radical so formed to hexafluoroethane and nitric oxide:

$$CF_3\cdot{}^* + CF_3\cdot NO \rightarrow (CF_3)_2NO\cdot \rightarrow C_2F_6 + NO \qquad (43)$$

Bis(trifluoromethyl) nitroxide is a stable compound [24] but, as formed from
an excited trifluoromethyl radical, is expected to decompose as indicated.
Trifluoronitrosomethane is known to be susceptible to radical attack on its
nitrogen.

The only other one-carbon compound studied is trichlorofluoromethane
(Freon 11) which was irradiated with an electron beam of 5 A at 1.5 MeV
from a Van de Graaff accelerator [25]. In the liquid phase at -70°C, and in
the presence of copper and stainless steel turnings, G values for the production of chlorine and fluorine of 0.62 and 0.16, respectively, were obtained.
The G values were determined by analysis of the deposit formed on the metal
turnings and the halogen found in fused potassium hydroxide, which was in
contact with the gas above the liquid trichlorofluoromethane. There was no
report of any organic products formed.

C. Synthetic Uses of Polyfluoromethanes

The most extensively studied reaction is the substitution of benzene by
intermediates derived from the γ radiolysis of bromotrifluoromethane [15].
Radiolysis of a 4:1 mixture of bromotrifluoromethane and benzene yields
fluorobenzene and benzotrifluoride.

The yield of fluorobenzene increased from G = 0.22 to 1.47 as the
temperature was raised from 27°C to 300°C, whereas the benzotrifluoride
yield first increased from G = 2.0 to 2.5 and then decreased to G = 1.9.
The formation of fluorobenzene was said to be due to radical reactions:

$$C_6H_6 + F\cdot \rightarrow C_6H_6F\cdot \qquad (44)$$
$$C_6H_6F\cdot + R\cdot \rightarrow C_6H_5F + RH \qquad (45)$$

The formation of benzotrifluoride may be ionic or free radical, and involves initially the addition of radical or ionic intermediates to the benzene, followed by abstraction of hydrogen by another reactive species:

$$C_6H_6 + CF_3 \cdot \quad \rightarrow \quad C_6H_6CF_3 \cdot \quad\quad (46)$$

$$C_6H_6CF_3 \cdot + R \cdot \quad \rightarrow \quad C_6H_5CF_3 + RH \quad\quad (47)$$

or

$$C_6H_6 + CF_3^+ \quad \rightarrow \quad C_6H_6CF_3^+ \quad\quad (48)$$

$$C_6H_6CF_3^+ + A^- \quad \rightarrow \quad C_6H_5CF_3 + HA \quad\quad (49)$$

Substitution on aromatic substrates has also been studied using tetrafluoromethane [16]. With benzene, fluorobenzene and benzotrifluoride are produced in comparable yields. The principal product from the irradiation of nitrobenzene with tetrafluoromethane is m-fluoronitrobenzene, although p-fluoronitrobenzene can be produced under favorable conditions. No 1-fluoronitrobenzene or trifluoromethylated products have been detected. Treatment of toluene in a similar way produces fluorotoluene and trifluoromethylated toluenes. In all three cases, the yield of products increases linearly with increasing electron fraction of tetrafluoromethane but is independent of the temperature of irradiation.

III. COMPOUNDS CONTAINING TWO CARBON ATOMS

A. Hexafluoroethane

The radiolysis of hexafluoroethane in the gas phase at 3 atm using a brass reaction vessel gives the following products (and 100 eV yields): tetrafluoromethane (1.6), hexafluorocyclopropane (0.30), octafluoropropane (0.21), decafluorobutane (0.14), and hexafluorodimethyl ether (0.03) [26]. The latter product presumably arises by reaction of oxygen adsorbed on the surface of the reaction vessel. Added oxygen increases the yield of hexafluorodimethyl ether to $G(CF_3OCF_3)$, $= 0.16$. In the presence of oxygen the yield of tetra-fluoromethane is halved, the yield of perfluorocyclopropane increases, and octafluoropropane and decafluorobutane are virtually eliminated. Table 3 shows details of the radiolysis results. Since the yields of octafluoropropane and decafluorobutane are totally scavenged by oxygen their formation is attributed to radical reactions

$$CF_3\cdot + C_2F_5\cdot \rightarrow C_3F_8 \tag{50}$$

$$2\ C_2F_5\cdot \rightarrow C_4F_{10} \tag{51}$$

Tetrafluoromethane on the other hand is only partly scavenged by oxygen and its formation is therefore attributed to both radical and ionic reactions

$$F\cdot + CF_3\cdot \rightarrow CF_4 \tag{52}$$

$$CF_3^+ + C_2F_6 \rightarrow CF_4 + C_2F_5^+ \tag{53}$$

$$C_2F_5^+ + C_2F_6 \rightarrow C_3F_7^+ + CF_4 \tag{54}$$

It appears to the authors that reaction (52) is unlikely and a more probable radical reaction is fluorine abstraction by an excited trifluoromethyl radical from a molecule of hexafluoroethane

$$CF_3\cdot^* + C_2F_6 \rightarrow CF_4 + C_2F_5\cdot \tag{55}$$

The chloropentafluoroethane detected in the radiolysis products is attributed to chlorotrifluoromethane, an impurity originally present in the hexafluoroethane.

A study of the radiolysis of hexafluoroethane in the presence of rare gases has been used to assess the energy transfer properties of the rare gases [27]. From a consideration of total product yields, it was shown that the rare gases have the following order of increasing ability to transfer energy to hexafluoroethane: He, Xe, Ne, Kr, Ar. This order demonstrates a correlation between the recombination energies of the rare gases and the expected ionization potential of hexafluoroethane. The ionization potential of hexafluoroethane is not accurately known, but it is probably similar to that of tetrafluoromethane and lies in the 14-16 eV range. The recombination energies of Kr^+ and Ar^+ are 14.4 and 15.8 eV, respectively. Values for the other rare gases lie well outside this range. It is therefore concluded that charge transfer makes an important contribution to the overall energy transfer process.

In the liquid-phase radiolysis of hexafluoroethane, as in the gas phase, oxygen reduces the yield of tetrafluoromethane and prevents the formation of octafluoropropane and decafluorobutane [28]. The oxygen-containing products, CO_2, CF_3OCF_3, and $C_2F_5OCF_3$, were also identified but not determined quantitatively. The presence of other oxygen-containing products was suspected.

Kevan et al. proposed a series of radical reactions for the radiolysis in the absence of oxygen:

$$CF_3\cdot + CF_3\cdot \rightarrow C_2F_6 \tag{56}$$

$$CF_3\cdot + C_2F_5\cdot \rightarrow C_3F_8 \tag{57}$$

$$C_2F_5\cdot + C_2F_5\cdot \rightarrow C_4F_{10} \tag{58}$$

$$CF_3\cdot + F\cdot \rightarrow CF_4 \tag{59}$$

$$C_2F_5\cdot + F\cdot \rightarrow C_2F_6 \tag{60}$$

$$F\cdot + F\cdot \rightarrow F_2 \tag{61}$$

$$CF_3\cdot + F_2 \rightarrow CF_4 + F\cdot \tag{62}$$

$$C_2F_5 + F_2 \rightarrow C_2F_6 + F\cdot \tag{63}$$

But, the existence of fluorine in either the atomic or molecular state is unlikely and hence reactions (59) to (63) are not favored. A sequence of reactions initiated by abstraction of fluorine from a molecule of hexafluoroethane by an excited trifluoromethyl radical explains the formation of the observed products and the stoichiometry:

$$4\ CF_3\cdot + 4\ C_2F_6 \rightarrow 4\ CF_4 + 4\ C_2F_5\cdot \tag{64}$$

$$2\ C_2F_5\cdot + 2\ CF_3\cdot \rightarrow 2\ C_3F_8 \tag{65}$$

$$C_2F_5\cdot + C_2F_5\cdot \rightarrow C_4F_{10} \tag{66}$$

Summing reactions (64), (65), and (66) gives overall:

$$6\ CF_3\cdot + 4\ C_2F_6 \rightarrow 4\ CF_4 + 2\ C_3F_8 + C_4F_{10} \tag{67}$$

and a product ratio of 4:2:1. This agrees, within the limits of experiment, with the observed ratio of 3.82:1.93:1.

TABLE 3

Results of the Radiolysis of Hexafluoroethane

Reactants	G values of products						F/C ratio	$-C_2F_6$
	CF_4	C_2F_5Cl	C_3F_8	C_4F_{10}	CF_3OCF_3	CO_2		
C_2F_6 gas, 3 atm, 7 Mrad	1.6	0.30	0.21	0.14	0.03	<0.01	3.02	1.9
C_2F_6 gas, 2 atm at 40°C	2.5	0.57	0.45	0.20			3.3	2.4
C_2F_6 gas + 0.5% O_2, 3 atm, 7 Mrad	0.80	<0.01	<0.01	<0.01	0.16	1.0	2.45[a]	
C_2F_6 gas + 1% O_2, 3 atm, 7 Mrad	0.84	0.59	<0.01	<0.01	0.16	1.1	2.46[a]	2.0
C_2F_6 gas + 1% O_2, 2 atm, 40°C	1.3		0.0	0.0				
C_2F_6 liquid, -78°C	1.72		0.87	0.45			3.0	3.0
C_2F_6 liquid, -78°C + nominal 10% O_2	0.97		0.0	0.0			3.0	~ 3
55% C_2F_6, 45% Xe	1.4[b]	0.21[b]	0.22[b]	0.15[b]	0.03[b]		3.01	1.7
35% C_2F_6, 65% Ar	1.8[b]	0.34[b]	0.46[b]	0.15[b]	0.06[b]		2.97	2.5
39.5% C_2F_6, 59.5% Ar, 1% O_2	1.3	0.58	0.24	<0.01	0.16	0.91	2.66[a]	2.5[c]
C_2F_6 gas, 99.5% He	3.9 (0.45)[b]		0.46 (0.053)[b]	0.19 (0.022)[b]				
C_2F_6 gas, 98% Ne	12.2 (1.25)[b]		2.39 (0.24)[b]	0.97 (0.10)[b]				

C_2F_6 gas, 98% Ar	34.1 (2.25)^b	6.86 (0.45)^b	2.30 (0.15)^b
C_2F_6 gas, 95% Kr	20.8 (2.14)^b	3.43 (0.35)^b	1.16 (0.12)^b
C_2F_6 gas, 96% Xe	22.7 (1.34)^b	1.51 (0.09)^b	0.50 (0.03)^b

^{a} Without CO_2.

^{b} Calculated on the basis of energy absorbed in hexafluoroethane only.

^{c} This assumes that all radicals reacting with oxygen are detected as carbon dioxide.

TABLE 4

Oxidation of 1:1 Mixtures of Tetrafluoroethylene and Oxygen

Initial pressure (mm Hg)	760	650	760	760
	x rays 0.4 Mrad/hr	γ rays 0.4 Mrad/hr	uv light	100°C
Compound	1.5 hr	1.5 hr	24 hr	14 hr
C_2F_4	25.3	22.0	64.2	15.5
O_2	18.3	15.7	43.0	8.5
COF_2	47.2	47.5	47.4	41.7
C_2F_4O	37.2	30.1	40.0	14.7
$(CF_2)_3$	4.5	5.7	7.0	42.0

$$C_2F_4 \rightarrow C_2^*F_4^*$$
$$C_2F_4^* \rightarrow 2R\cdot$$
$$R\cdot + O_2 \rightarrow RO_2\cdot$$
Initiation

thus $RO_2\cdot + C_2F_4 \rightarrow R-O\cdot O\cdot C_2F_4\cdot$, etc.   Copolymerization

Termination processes may be first or second order.

$$R'O_2\cdot \rightarrow \text{peroxidic liquid}$$
$$R'O_2\cdot + R'O_2\cdot \rightarrow \text{peroxidic liquid}$$
$$R'O\cdot + R'O\cdot \rightarrow \text{peroxidic liquid}$$
$$R'O_2\cdot + R'O\cdot \rightarrow \text{peroxidic liquid}$$

where R' represents a copolymer chain.

Carbonyl fluoride and tetrafluoroethylene oxide are thought to be formed by bimolecular oxygen atom transfer and polymer chain degradation

$$R'O_2\cdot + C_2F_4 \rightarrow R'\cdot O\cdot + (CF_2)_2O + COF_2$$
$$\rightarrow R'\cdot + (CF_2)_2O + COF_2$$

This conjoint formation of carbonyl fluoride and tetrafluoroethylene oxide is supported by the kinetic data. The different kinetic trend for the formation of perfluorocyclopropane suggests that the peroxidic intermediate is not involved in this case.

B. Tetrafluoroethylene

The radiolytic oxidation of tetrafluoroethylene is one of the most interesting reactions reported recently [29-31] . When a 1:1 mixture of oxygen and
tetrafluoroethylene is irradiated with γ rays at 0.4 Mrad/hr, high yields of
carbonyl fluoride, tetrafluoroethylene oxide, and a polyperoxidic liquid are
produced together with a small amount of hexafluorocyclopropane. The products account for more than 90% of the gaseous mixture. G values of the order
1000 - 3000 are observed, indicating that the reaction proceeds by a chain
mechanism. The radiation results are compared with the photochemical and
thermal oxidation results in Table 4.

A detailed kinetic study of this oxidation shows that carbonyl fluoride
and tetrafluoroethylene oxide have the same apparent activation energy of
4.3 kcal/mole. A value of 6.7 kcal/mole is obtained for perfluorocyclopropane. The overall rates of formation of carbonyl fluoride, tetrafluoroethylene oxide, and perfluorocyclohexane are $(0.26 \pm 0.02) I^{0.72} \{C_2 F_4\}$,
$(0.117 \pm 0.006) I^{0.72} C_2 F_4$, and $(0.78 \pm 0.03) 10^{-2} I^{0.5} \{C_2 F_4\}$ respectively,
all in mole/$l^{-1} \cdot h^{-1} \times 10^{3*}$.

The NMR spectrum of the polyperoxide indicates that only CF_2 groups
are bound to oxygen atoms. It is therefore formulated as a 1:1 copolymer of
tetrafluoroethylene and oxygen--poly(tetrafluoroethyleneoxide) (PTFEO).
PTFEO has also been made by the irradiation of tetrafluoroethylene oxide
at −130°C [32]. An ESR study of the γ-irradiated PTFEO also supports the
formulation as a 1:1 copolymer [33]. Thus, spectra consistent with the
formation of the radicals $-CF_2 \cdot CF_2 \cdot CF_2 \cdot$ and $-O \cdot \overset{.}{C}F \cdot CF_2 \cdot O$ − were observed
from a sample of the irradiated copolymer in vacuo.

The copolymerization of tetrafluoroethylene and oxygen by ionizing radiation is considered to be initiated by radical species derived from tetrafluoroethylene:

* The dose rate (I) is expressed in megaroentgen h^{-1}.

C. Other Fluoroolefins

The radiolytic oxidation of perfluoropropene has been briefly reported. The products are carbonyl fluoride, trifluoroacetyl fluoride, and a material with a high oxygen content [34].

Thermal and photochemical addition of various species across fluoroolefins is well known; this reaction can also be initiated by γ radiation [35–37]. Thus tetrahydrofuran, 1,4-dioxane, diethyl ether, and various alcohols form 1:1 adducts with the olefins $CFCl = CCl_2$, $CFCl = CFCl$, $CF_2 = CCl_2$; for example,

In addition, diethyl ether also forms a 1:2 adduct

$$CH_3CH_2OCH_2CH_3 + CFX = CX_2 \;\lambda \to CH_3CH_2O\,CH(CFX\;CHX_2)CH_3$$
$$+ CH_3CH(CFX\;CHX_2)OCH(CFX\;CHX_2)CH_3$$

IV. ALIPHATIC COMPOUNDS
CONTAINING THREE OR MORE CARBON ATOMS

Perfluoroalkanes ranging from C_3 to C_8 have been studied [14]. Aluminum vessels were used but the formation of small amounts of oxygen-containing products indicates that no attempt was made to prefluorinate the surface of the reaction vessels. A wide range of fluorocarbon products were formed ranging from CF_4 to $C_{12}F_{26}$. In all cases products of higher molecular weight than the starting material were observed and a preponderance of branched product molecules containing trifluoromethyl groups was noted. The more branched the parent molecule, the higher is the yield of carbon tetrafluoride, for example, $n\text{-}C_6F_{14}$ shows a $G(CF_4) = 0.59$, whereas $2,3\text{-}(CF_3)_2C_4F_8$ shows a $G(CF_4) = 2.02$.

The following conclusions may be drawn:

1. C–F bonds at secondary and tertiary carbon atoms are more susceptible to radiation damage than those at trifluoromethyl groups.

2. Trifluoromethyl groups, once formed, do not dissociate further but either abstract fluorine or pick up a fluorine atom to form carbon tetrafluoride.

3. Radical–radical combination reactions occur frequently.

Taking perfluoropentane as a typical example, the following reaction scheme is illustrative:

$$\begin{aligned}
&&\rightarrow\quad & CF_3{}^{\cdot} \;+\; CF_3CF_2CF_2CF_2{}^{\cdot}\\
&&\rightarrow\quad & C_2F_5{}^{\cdot} \;+\; CF_3CF_2CF_2{}^{\cdot}\\
n\text{-}C_5F_{12} &&\rightarrow\quad & F^{\cdot} \;+\; CF_3(CF_2)_3CF_2{}^{\cdot}\\
&&\rightarrow\quad & F^{\cdot} \;+\; CF_3\overset{\cdot}{C}FCF_2CF_2CF_3\\
&&\rightarrow\quad & F^{\cdot} \;+\; (C_2F_5)_2\overset{\cdot}{C}F
\end{aligned}$$

Such reactions are followed by the radical–radical recombination reactions

$$\begin{aligned}
CF_3CF_2CF_2CF_2{}^{\cdot} \;+\; CF_3CF_2CF_2{}^{\cdot} &\rightarrow n\text{-}C_7F_{16}\\
CF_3(CF_2)_3CF_2{}^{\cdot} \;+\; C_2F_5{}^{\cdot} &\rightarrow n\text{-}C_7F_{16}\\
CF_3\overset{\cdot}{C}FCF_2CF_2CF_3 \;+\; C_2F_5{}^{\cdot} &\rightarrow 3\text{-}(CF_3)C_6F_{13}\\
(C_2F_5)_2CF^{\cdot} \;+\; C_2F_5{}^{\cdot} &\rightarrow 3\text{-}(C_2F_5)C_5F_{11}
\end{aligned}$$

V. ALICYCLIC COMPOUNDS

Perfluorocyclohexane has received the most study, both at high and low dose [38, 39]. Perfluorobicyclohexyl is the predominant product, but the yield decreases with increasing dose. Other C_{12} products formed in low yield include perfluoromethylcyclohexane and perfluoroethylcyclohexane. The low-molecular-weight compounds, CF_4, C_2F_6, and C_3F_8, were also identified with a total G value of less than 0.3.

Perfluorodecalin [39], perfluorobicyclohexyl [39], and perfluorodimethyl–cyclohexane [40] behave similarly upon radiolysis, in that gases in the C_1 - C_4 range are produced together with higher–molecular–weight material.

The radiolysis of perfluorocyclobutane has received more detailed study [38, 38a]. The products observed together with respective G values are C_2F_4 (0.13), C_3F_8 (0.008), C_3F_6 (0.005), C_5F_{12} (0.065), C_6F_{12} (0.091), C_7F_{14} (0.077), C_8F_{16} (0.21), C_9F_{18} (0.13), C_9F_{20} (0.007), $C_{10}F_{20}$ and $C_{10}F_{18}$ (0.12), C_{12}, C_{13}, and C_{14} fluorocarbons (0.47), and a white polymer. The occurrence of a polymer, presumably of the PTFE type, suggests that fluorocarbons greater than C_{14} may also be formed. The product distribution and yields were unaffected by pressure or the addition of 5% nitrous oxide. The latter observations suggests that perfluorocyclobutane is a better electron scavenger than nitrous oxide.

The primary processes are considered to be

$$c\ C_4F_8\ \rightsquigarrow\ c\text{-}C_4F_7\cdot\ +\ F\cdot$$
$$\rightsquigarrow\ 2C_2F_4$$
$$\rightsquigarrow\ CF_3\cdot\ +\ CF_2\text{=}CF\text{-}CF_2\cdot$$

although intermediate ionic species are not discounted. The formation of a white polymer is attributed to the polymerization of tetrafluoroethylene by radicals or atoms formed in the primary processes, for example,

$$
\begin{array}{c}
CF_2\text{-}CF\cdot \\
|\quad\ \ | \\
CF_2\text{-}CF_2
\end{array}
\xrightarrow{C_2F_4}
\begin{array}{c}
CF_2\text{-}CF\text{-}CF_2\text{-}CF_2\cdot \\
|\qquad\qquad\ | \\
CF_2\text{-}CF_2
\end{array}
\xrightarrow{C_2F_4}
\ \text{etc.}
$$

Chain termination results from radical–radical reactions, such reactions giving rise to the other observed products, for example,

$$
C_3F_7\cdot\ +\
\begin{array}{c}
CF_2\text{-}CF_2 \\
|\quad\ \ | \\
CF_2\text{-}CF\cdot
\end{array}
\rightarrow
\begin{array}{c}
CF_2\text{-}CF_2 \\
|\quad\ \ | \\
CF_2\text{-}CF\cdot C_3F_7
\end{array}
$$

VI. AROMATIC COMPOUNDS

The compounds hexafluorobenzene [39, 41-43], perfluorobiphenyl [39], perfluoronaphthalene [39], and perfluoro-o-terphenyl [41] have been studied. Their behavior is typified by the very low gas yields and the formation of a so-called "polymer." For example, the irradiation of hexafluorobenzene in glass or nickel vessels gives rise to a polymer with a yield of G(polymer) ~ 2 and an average molecular weight of about 1200. The structure of this polymer or that obtained from any other aromatic molecule has not been determined. A more careful examination of the products arising from the radiolysis of hexafluorobenzene has demonstrated the formation of hexafluorobicyclo{2.2.0} hexa-2,5-diene (Dewar C_6F_6). A similar type of isomerization has been observed during the photolysis of hexafluorobenzene. It is suggested that isomer formation occurs from low-lying excited states. These excited states may be formed by direct excitation, by ion-electron neutralization, or by radical-fluorine recombination.

VII. ELECTRON CAPTURE AND
PERFLUOROCARBONS AS ELECTRON SCAVENGERS

The initial absorption of γ rays, as stated earlier, can give rise to ionization with the production of positive ions and electrons. Electrons produced by the γ irradiation of a substrate are generally energetic enough to cause further ionization, but in doing so they are eventually slowed down to thermal energies. Such electrons are usually referred to as thermal electrons. The number of thermal electrons produced by the absorption of a quantum of γ radiation is high, and the subsequent fate of these electrons has a marked effect on the radiation-chemistry process.

An obvious fate of thermal electrons is direct neutralization with a positive ion to give an excited molecule. A less obvious fate is capture of the thermal electron by a neutral molecule to give a negative ion, which may itself then be neutralized by a positive ion. Electron capture by neutral molecules has been demonstrated experimentally by Hamill and his colleagues by irradiating frozen solutions of biphenyl or naphthalene in organic solvents [44]. The species $C_{12}H_{10}^{-}$ and $C_{10}H_{8}^{-}$ have been detected spectroscopically in these solutions. Further work by Hamill involving the use of a second solute which could compete with the biphenyl or naphthalene for electrons produced a decrease in $G(C_{12}H_{10}^{-})$ or $G(C_{10}H_{8}^{-})$ [45]. Compounds which decreased these G values included iodine, carbon tetrabromide, sulfur dioxide, carbon tetrachloride, and nitrous oxide. Other compounds which have subsequently been found to capture electrons are sulfur hexafluoride and perfluorocarbons. These compounds are termed electron scavengers. The use and mode of operation of perfluorocarbons as electron scavengers merits discussion here.

The radiolysis of hydrocarbons in the presence of electron scavengers results in a reduction of the yield of hydrogen relative to that in the absence of scavengers. This is attributed to the capture of electrons which would otherwise interact with a positive ion and eventually produce hydrogen. The introduction of electron scavengers into a system is therefore useful in attempting to establish the part which ionic reactions play in the radiolysis process [46 - 48].

Dyne has postulated a mechanism for the mode of operation of electron scavengers during the radiolysis of hydrocarbons [49], and it has received general support from other workers. The radiolysis of a hydrocarbon A produces initially positive ions, electrons, and excited molecules.

$$A \rightsquigarrow A^{+} + e^{-} \tag{68}$$
$$\rightsquigarrow A^{*} \tag{69}$$

Recapture of the electron by a positive ion can produce an excited species
which is the same as A^* above.

$$A^+ + e^- \rightarrow A^* \tag{70}$$

However, an electron scavenger S may also react rapidly with electrons to
give negative ions

$$S + e^- \rightarrow S^- \tag{71}$$

These scavenger anions are then neutralized by reaction with positive ions;
however, this does not produce the same active species A^* as does direct
neutralization with an electron

$$S^- + A^+ \rightarrow S + \text{"A" (not } A^*) \tag{72}$$

In the case of hydrocarbons, A^* can break down as follows to produce hydrogen:

$$X_H + \text{products} \tag{73}$$

$$A^* \big\langle$$

$$H_2 + \text{products} \tag{74}$$

$$X_H + A \rightarrow H_2 + \text{products} \tag{75}$$

Therefore, reaction of an electron scavenger with electrons will reduce the
hydrogen yield by reduction of reaction (70).

The fact that the excited state A^* is not formed in reaction (72), does not
mean that breakdown cannot occur on neutralization. It is postulated that the
nature of the electron scavenger determines whether dissociation follows
neutralization or not. The process is referred to as dissociative charge
transfer or nondissociative charge transfer. The yield of other products
from hydrocarbon radiolysis is also altered in the presence of electron
scavengers, and this is attributed to secondary reactions of the anion S^- on
neutralization [50-52].

The occurrence of dissociative and nondissociative charge transfer
processes is demonstrated by the effect of nitrous oxide and sulfur hexafluoride on the radiolysis of liquid cyclohexane [50]. Both nitrous oxide and
sulfur hexafluoride reduce the yield of hydrogen to about one-half of its
value in pure cyclohexane, showing that electron scavenging is taking place

for both solutes. However, nitrous oxide is found to increase the yield of cyclohexene and bicyclohexyl, whereas sulfur hexafluoride reduces these yields to about one-half of the values obtained for pure cyclohexane. In the case of nitrous oxide, nitrogen is produced together with a small amount of cyclohexanol showing that breakdown of nitrous oxide occurs. The explanation is that, for nitrous oxide, dissociative charge transfer can occur, whereas for sulfur hexafluoride, it cannot. The mechanism for nitrous oxide is:

$$C_6H_{12} \xrightarrow{} C_6H_{12}^+ + e^- \tag{76}$$

$$\xrightarrow{} C_6H_{12}^* \tag{77}$$

$$C_6H_{12}^+ + e^- \rightarrow C_6H_{12}^* \tag{78}$$

$$C_6H_{12}^* \rightarrow H\cdot + C_6H_{11}\cdot \tag{79}$$

$$\rightarrow H_2 + C_6H_{10} \tag{80}$$

$$H\cdot + C_6H_{12} \rightarrow H_2 + C_6H_{11}\cdot \tag{81}$$

$$e^- + N_2O \rightarrow \{N_2O^-\} \rightarrow N_2 + O^- \tag{82}$$

$$O^- + C_6H_{12}^+ \rightarrow \text{''}C_6H_{12}\text{''} \ (\text{not } C_6H_{12}^*) + O \tag{83}$$

Since bicyclohexyl and cyclohexene yields are not reduced when nitrous oxide is added, reaction (83) must give intermediates which yield cyclohexyl radicals. Two possibilities exist [51, 52]:

$$O^- + C_6H_{12}^+ \rightarrow OH\cdot + C_6H_{11}\cdot \tag{84}$$

$$OH\cdot + C_6H_{12} \rightarrow H_2O + C_6H_{11}\cdot \tag{85}$$

or

$$O^- + C_6H_{12} \rightarrow OH^- + C_6H_{11}\cdot \tag{86}$$

$$OH^- + C_6H_{12}^+ \rightarrow H_2O + C_6H_{11}\cdot \tag{87}$$

This is dissociative charge transfer.

For sulfur hexafluoride the mechanism is

$$SF_6 + e^- \rightarrow SF_6^- \tag{88}$$

$$SF_6^- + C_6H_{12}^+ \rightarrow \text{''}C_6H_{12}\text{''} \ (\text{not } C_6H_{12}^*) \tag{89}$$

The products of "C_6H_{12}" are not cyclohexyl radicals, i.e., nondissociative charge transfer.

The possible use of perfluorocarbons as electron scavengers was first realized by Rajbenbach [53]. He noted that several perfluorocarbons gave a high response on the electron-capture detector [54], and concluded that they should have a high reactivity towards thermal electrons produced during radiolysis. He studied the radiolysis of n-hexane and cyclohexane in the presence of dilute solutions of perfluoromethane, perfluoroethane, perfluoro-cyclobutane, perfluorocyclohexane, perfluoromethylcyclohexane, and per-fluorobenzene. All these fluorocarbons, with the exception of perfluoromethane and perfluoroethane, markedly reduce the hydrogen yields from both n-hexane and cyclohexane. Electron capture was thought to be responsible for this effect. Thermal hydrogen atom scavenging by the perfluorocarbon solute would not seem to be a plausible alternative because of the high strength of the C-F bond. The ionization potentials of fluorocarbons are usually greater than those of corresponding hydrocarbons; therefore charge transfer from solvent to solute cannot occur either.

Confirmation that the fluorocarbons were scavenging electrons was obtained from a study of the radiolysis of n-hexane and cyclohexane in the presence of a combination of nitrous oxide and a fluorocarbon. Nitrous oxide is known to react with electrons formed in the radiolysis of cyclohexane to produce nitrogen [46]. A second solute competing with nitrous oxide for electrons should therefore reduce the yield of nitrogen, $G(N_2)$. This was demonstrated in the case of the cyclic fluorocarbons but not in the case of perfluoromethane or perfluoroethane. The observation that perfluoromethane did not compete with nitrous oxide for thermal electrons was consistent with the finding that perfluoromethane forms no negative ion when bombarded with low-energy electrons [55].

The radiolysis of cyclohexane in the presence of scavenger quantities of perfluorocyclohexane, perfluorocyclobutane, perfluoromethylcyclohexane, and perfluorobenzene has been extensively studied by Sagert and co-workers, who present evidence favoring a dissociative charge transfer process [56-58]. In all three cases the yield of hydrogen was reduced in the presence of a fluorocarbon scavenger. The yields of the products cyclohexene and

bicyclohexyl were also affected, decreasing in the presence of perfluoromethylcyclohexane and perfluorocyclobutane, but increasing in the presence
of perfluorocyclohexane. In addition, large yields of undecafluorocyclohexane
and heptafluorocyclobutane were produced indicating that C-F bond cleavage
had occurred. It was demonstrated that electron scavenging preceded undecafluorocyclohexane formation by introducing a second electron scavenger —
nitric oxide — into the system. This reduced the yield of undecafluorocyclohexane while, at the same time nitrogen was produced. Further work
showed that perfluorocyclohexane is twice as efficient an electron scavenger
in this system as nitrous oxide. These results were interpreted in terms of
electron scavenging followed by a dissociative charge transfer process.

$$C_6F_{12} + e^- \quad \rightarrow \quad C_6F_{12}^- \tag{90}$$

$$C_6F_{12}^- \quad \rightarrow \quad C_6F_{11}\cdot + F^- \tag{91}$$

$$F^- + C_6H_{12}^+ \quad \rightarrow \quad HF + C_6H_{11}\cdot \tag{92}$$

$$C_6F_{11}\cdot + C_6H_{12} \quad \rightarrow \quad C_6F_{11}H + C_6H_{11}\cdot \tag{93}$$

Reaction (91) is extremely endothermic in the gas phase and therefore unlikely unless the energetics in the liquid phase are markedly different from
those in the gas phase [59].

Another possibility is

$$C_6F_{12} + C_6H_{12}^+ \rightarrow C_6F_{11}\cdot + C_6H_{11}\cdot + HF \tag{94}$$

followed by reaction (93). Since the anion $C_6F_{12}^-$ has a long lifetime this
proposal is an attractive one [60, 61].

In order to distinguish between these two mechanisms, a study in the gas
phase was made, where reaction (91) may be ruled out. The addition of
perfluorocyclohexane decreased the hydrogen yield to $G(H_2) = 3.0$, increased
the yields of cyclohexane and bicyclohexyl, and produced undecafluorocyclohexane, $G(C_6F_{11}H) = 2.0$. It may be calculated that the electron yield for
cyclohexane is $G(e^-) = 4.2$ [19, 20]. Therefore, if all the electrons are
scavenged by perfluorocyclohexane, then $G(C_6F_{12}^-)$ should equal 4.2;
assuming reaction (91) does not occur, $G(C_6F_{11}\cdot)$ and $G(C_6F_{11}H)$ should
also equal 4.2. This neglects the formation of undecafluorocyclohexyl
(cyclohexane), which was detected in small yield. Therefore, since the

yield of undecafluorocyclohexane was only about 2, the production of the radical C_6F_{11}, must have occurred only for half the neutralizations in the gas phase. Therefore, reaction (94) must proceed at least to some extent, in the gas phase, although no conclusions could be drawn regarding the source of $C_6F_{11} \cdot$ radicals in the liquid phase.

In the solid state, the irradiation of cyclohexane in the presence of per-fluorocyclohexane gives only a small yield of undecafluorocyclohexane, $G(C_6F_{11}) \sim 0.3$. It appears that, in the solid state, neutralization of anion and cation occurs without bond breakage.

REFERENCES

[1] S.C. Lind, Monatsh Chem., 33, 295 (1912).

[2] H. Eyring, J.O. Hirschfelder, and H.S. Taylor, J. Chem. Phys., 4, 479, 570 (1936).

[3] P. Harteck and S. Dondes, Nucleonics, 14, 66 (1956).

[4] H.A.J.B. Battaerd and G.W. Tregear, Rev. Pure Appl. Chem., 16, 83 (1966).

[5] H.A. Bethe, Z. Physik, 76, 293 (1932).

[6] A.H. Samuel and J.L. Magee, J. Chem. Phys., 21, 1080 (1953).

[7] G.G. Meisels, W.H. Hamill, and R.R. Williams, J. Phys. Chem., 61, 1456 (1957).

[8] L.M. Dorfman and M.C. Sauer, J. Chem. Phys., 32, 1886 (1960).

[9] R.H. Schuler, J. Phys. Chem., 62, 37 (1958).

[10] K. Yang and P.J. Manno, J. Am. Chem. Soc., 81, 3507 (1959).

[11] K. Yang, J. Phys. Chem., 65, 42 (1961).

[12] J. Fajer, D.R. MacKenzie, and F.W. Bloch, J. Phys. Chem., 70, 935 (1966).

[13] W.C. Askew and T.M. Reed, Nucl. Sci. Eng., 29, 143 (1967).

[14] W.C. Askew, T.M. Reed, and J.C. Mailen, Radiation Res., 33, 282 (1968).

[15] P.Y. Feng, U. S. Atomic Energy Commission Rep. No. COO-578-3-32, 1966.

[16] P.Y. Feng, W.A. Glasson, L. Mamula, K. Schmude, and S. Soffer, Proc. Japan Conf. Radioisotopes, 4, 445 (1961).

[17] R.E. Florin, D.W. Brown, and L.A. Wall, J. Phys. Chem., 66, 2672 (1962).

[18] R.W. Fessenden and R.H. Schuler, J. Chem. Phys., 43, 2704 (1965).

[19] P.B. Ayscough and E.W.R. Steacie, Proc. Roy. Soc. (London), Ser. A, 234, 476 (1956).

[20] J.R. Dacey and D.M. Young, J. Chem. Phys., 23, 1302 (1955).

[21] I. McAlpine and H. Sutcliffe, J. Phys. Chem., 73, 3215 (1969).

[22] I. McAlpine and H. Sutcliffe, J. Phys. Chem., 74, 848 (1970).

[23] I. McAlpine and H. Sutcliffe, J. Phys. Chem., 74, 1422 (1970).

[24] W.D. Blackley and R.R. Reinhard, J. Am. Chem. Soc., 87, 802 (1965).

[25] M.D. Silverman, B.O. Heston, and P.S. Rudolph, Nucl. Sci. Eng., 15, 217 (1963).

[26] L. Kevan and P. Hamlet, J. Chem. Phys., 42, 2255 (1965).

[27] L. Kevan, J. Chem. Phys., 44, 683 (1966).

[28] A. Sokolowska and L. Kevan, J. Phys. Chem., 71, 2220 (1967).

[29] V. Caglioti, M. Lenzi, and A. Mele, Nature, 201, 610 (1964).

[30] V. Caglioti, A. Delle Site, M. Lenzi, and A. Mele, J. Chem. Soc., 1964, 5430 (1964).

[31] D. Cordischi, M. Lenzi, and A. Mele, Trans. Faraday Soc., 60, 2047 (1964).

[32] P. Barnaba, D. Cordischi, M. Lenzi, and A. Mele, Chim. Ind. (Milan), 47, 1060 (1965).

[33] P. Barnaba, D. Cordischi, A. Delle Site, and A. Mele, J. Chem. Phys., 44, 3672 (1966).

[34] M. Lenzi and A. Mele, Nature, 205, 1104 (1965).

[35] H. Muramatsu, K. Inukai, and T. Ueda, J. Org. Chem., 29, 2220 (1964).

[36] H. Muramatsu and K. Inukai, J. Org. Chem., 30, 544 (1965).

[37] H. Muramatsu, J. Org. Chem., 27, 2325 (1962).

[38] M.B. Fallgatter and R.J. Hanrahan, J. Phys. Chem., 69, 2059 (1965).

[38a] E. Heckel and R.J. Hanrahan, Advances in Chemistry Series, No. 82, Radiation Chemistry II, 120, 1968.

[39] D.R. MacKenzie, F.W. Bloch, and R.H. Wiswall, J. Phys. Chem., 69, 2526 (1965).

[40] T.H. Scott and J.A. Wethington, Nucl. Sci. Eng., 38, 48 (1969).

[41] R.E. Florin, L.A. Wall, and D.W. Brown, J. Res. Natl. Bur. Std., 64A, 269 (1960).

[42] F.W. Bloch and D.R. MacKenzie, J. Phys. Chem., 73, 552 (1969).

[43] J. Fajer and D.R. MacKenzie, J. Phys. Chem., 71, 784 (1967).

[44] P.S. Rao, J.R. Nash, J.P. Guarino, M.R. Ronayne, and W.H. Hamill, J. Am. Chem. Soc., 84, 500 (1962).

[45] J.P. Guarino, M.R. Ronayne, and W.H. Hamill, Radiation Res., 17, 379 (1962).

[46] G. Scholes and M. Simic, Nature, 202, 895 (1964).

[47] L.J. Forrestal and W.H. Hamill, J. Am. Chem. Soc., 83, 1535 (1961).

[48] J.A. Stone and P.J. Dyne, Radiation Res., 17, 353 (1962).

[49] P.J. Dyne, Can. J. Chem., 43, 1080 (1965).

[50] N.H. Sagert and A.S. Blair, Can. J. Chem., 45, 1351 (1967).

[51] S. Sato, R. Yugeta, K. Shinsaka, and T. Terao, Bull. Chem. Soc. Japan, 39, 156 (1966).

[52] R. Blackburn and A. Charlesby, Nature, 210, 1036 (1966).

[53] L.A. Rajbenbach, J. Am. Chem. Soc., 88, 4275 (1966); J. Chem. Phys., 47, 242 (1967).

[54] A.A. Altschuler and C.A. Clemons, Anal. Chem., 38, 133 (1966).

[55] V.M. Hickam and D. Berg, J. Chem. Phys., 29, 517 (1958).

[56] N.H. Sagert, Can. J. Chem., 46, 95 (1968).

[57] N.H. Sagert and A.S. Blair, Can. J. Chem., 46, 3284 (1968).

[58] N.H. Sagert, J.A. Reid, and R.W. Robinson, Can. J. Chem., 47, 2655 (1969).

[59] M.M. Bibby and G. Carter, Trans. Faraday Soc., 62, 2637 (1966).

[60] B.H. Mahan and C.F. Young, J. Chem. Phys., 44, 2192 (1966).

[61] R.K. Assundi and J.D. Craggs, Proc. Phys. Soc. (London), 83, 611 (1964).

Chapter 2

FLUORO-β-DIKETONES
AND METAL FLUORO-β-DIKETONATES

Paul Mushak,* Mary T. Glenn, and John Savory

Department of Pathology
University of Florida College of Medicine
Gainesville, Florida

* Present address: Department of Molecular Biophysics and Biochemistry, Yale University, New Haven, Connecticut.

I. INTRODUCTION

β-Diketones are of interest to workers in many different areas of chemistry, and the importance of the fluorine-substituted derivatives has been well
recognized.

A variety of structures otherwise synthetically inaccessible have been
prepared from fluoro-β-diketones, while the effects of fluorine sustituents
upon the chemical reactivity of β-diketones has prompted a number of
physical chemical studies.   These fluoro-β-diketones have proved of considerable interest and value to inorganic structural chemists studying the
corresponding transition metal derivatives.   In addition, utilization of fluoro-
β-diketones as chelation-extraction agents in a variety of analytical manipu-
lations has been extremely important.

This present review covers the literature pertaining to fluoro-β-diketones
through December 1968, and includes all pertinent aspects of the subject.
The metal analogs of the fluoro-β-diketones also are reviewed.

II. GENERAL CONSIDERATIONS
OF FLUORO-β-DIKETONES

Compounds in which a methylene group or a substituted carbon atom is
interposed between acyl or aroyl groups are designated as β-diketones.
Structurally, these agents are depicted as keto-enols, possessing a cis
configuration and a syn(cisoid) conformation permitting added stabilization
to be achieved via intramolecular hydrogen bonding [1-8].

The enol content in this series varies from about 80% for pentane-2,4-
dione [6] to 92-100% for the fluorine substituted derivatives [7,8].   In the
case of the monothio derivatives, the thiocarbonyl function appears in the
enolic form [1-5].

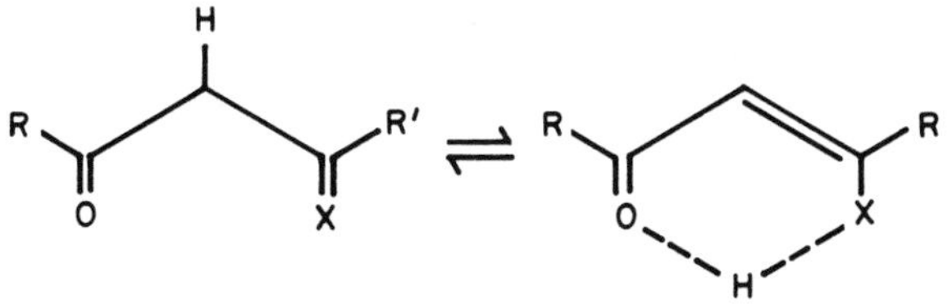

X = O, S
R = R' = alkyl = aryl
R = alkyl, R' = aryl

A. Synthesis of Fluoro-β-Diketones

Preparative procedures for the fluorine-containing β-diketones essentially
parallel those employed for the nonfluorinated compounds with some experimental variation. Syntheses of the nonfluorinated β-diketones have been
reviewed in detail [9].

1. Claisen Condensation Reaction

Fluoro-β-diketones usually are prepared by the base-promoted condensation of a fluorinated ester with a ketone containing at least one alpha
hydrogen atom. Sodium alkoxide [7, 8, 10-16], sodium amide [12], or
sodium hydride [17] have been employed as condensing agents with benzene
or ethyl ether being the solvents used for these reactions.

The condensation product can be isolated and purified via several procedures which include extraction, distillation, or separation using the
copper (II) chelate which later can be decomposed by acid to liberate the
β-diketone. Yields of the diketones using the Claisen condensation reaction
are optimal where the ketonic reactant possesses a methyl group. Increasing
substitution at the alpha position in the starting ketone results in reduced
yields of the fluoro-β-diketone due to the incursion of cleavage reactions [15].

$$R-\overset{O}{\overset{||}{C}}-CHR_1R_2 \ + \ C_nH_mF_{(2n+1)-m}-\overset{O}{\overset{||}{C}}-O-R_3 \ \xrightarrow{\ B^- \ } \ R-\overset{O}{\overset{||}{C}}-CR_1R_2-\overset{O}{\overset{||}{C}}-C_nH_mR_{(2n+1)-m}$$

$$R = alkyl, \ aryl$$
$$R_1, \ R_2 = H, \ alkyl, \ aryl$$
$$R_3 = C_2H_5$$
$$n = 1, \ 2, \ . \ . \ . \ .$$
$$m = 0, \ 1, \ . \ . \ . \ . \ .$$
$$B^- = OR^-, \ NH_2^-, \ H^-$$

2. Reaction of a Fluoroketone with an Acid Anhydride

Acid-catalyzed acylation of a ketone may be employed, usually involving boron trifluoride as a catalyst with an acid anhydride as the acylating agent.

$$C_nH_mF_{(2n+1)-m}-\overset{O}{\overset{||}{C}}-CH_3 \ + \ (R-CO)_2O \ \xrightarrow{\ BF_3 \ } \ C_nH_mR_{(2n+1)-m}-\overset{O}{\overset{||}{C}}-CH_2-\overset{O}{\overset{||}{C}}-R$$
$$+ \ R-COOH$$

$$R = alkyl, \ aryl$$
$$n = 1, \ 2, \ . \ . \ .$$
$$m = 0, \ 1, \ . \ . \ .$$

Thus, treatment of fluoroacetone with acetic anhydride in acetic acid saturated with boron trifluoride at low temperature results in the preparation of 1-fluoropentane-2, 4-dione [18].

3. Reaction of a β-Diketone with Perchloryl Fluoride

A valuable method for introduction of fluorine into the gamma position of a β-diketone involves the action of perchloryl fluoride on the sodium salt of the diketone [19].

$$R-\overset{O}{\overset{\|}{C}}-CHR''-\overset{O}{\overset{\|}{C}}-R' \xrightarrow{B^-} R-\overset{O}{\overset{\|}{C}} \diagdown \overset{\overset{-}{O}}{\diagup} -R' \xrightarrow{ClO_3F} R-\overset{O}{\overset{\|}{C}}-CFR''-\overset{O}{\overset{\|}{C}}-R'$$

R = R' = alkyl, aryl
R'= alkyl, H
B$^-$ = CH$_3$O$^-$, C$_2$H$_5$O$^-$

4. Preparation of Thio-β-Diketones

Thio-β-diketones are prepared directly from the oxo precursor using anhydrous hydrogen sulfide and hydrogen chloride at low temperatures [1-5].

$$R-\overset{O}{\overset{\|}{C}}-CHR''-\overset{O}{\overset{\|}{C}}-R' \xrightarrow[-20°]{H_2S \quad HCl} R-\overset{O\cdots H}{\overset{\diagup}{C}} \diagdown \overset{S}{\underset{R''}{C}}=\overset{}{C}-R'$$

R = CF$_3$, C$_2$H$_5$, . . .
R'= alkyl, thienyl, aryl
R''= H, alkyl

B. Chemical Reactivity

Substitution of fluorine for hydrogen in β-diketones results in a marked increase in the acidity of the methylene hydrogen(s) as shown in Table 1. The strong electron-withdrawing effect of the fluorine accounts for the increased acidity.

The introduction of fluorine enhances the reactivity of the carbonyl functions in β-diketones to nucleophilic agents, particularly water and the lower alcohols. Hexafluoroacetylacetone [H(hfa)] reacts with water and methanol to yield the corresponding tetraol [23] and dihemiketal [24], respectively. With ethylene glycol, H(hfa) reacts to give the dioxaheterocycle (1) [25].

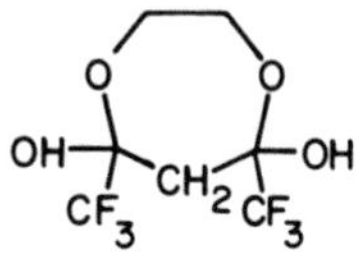

(1)

TABLE 1

Effect of the Introduction of Fluorine
on the Acidity of the Pentane-2,4-diones

β-Diketone	pKa	Ref.
Pentane-2,4-dione	8.9	[20]
1,1,1-Trifluoro-pentane-2,4-dione	6.7	[21]
1,1,1,5,5,5-Hexafluoropentane-2,4-dione	4.6	[22]

In aqueous solution, trifluoroacetylacetone [H(tfa)] forms the hydrate with water and the hemiketal with lower aliphatic alcohols [26]. Thenoyltrifluoroacetone [H(tta)] forms a hydrate in aqueous media, which is apparently more liable to cleavage than the unhydrated species [27]. At high pH values, the hydrate undergoes cleavage rather than enolization.

In the presence of strong acid systems, fluorinated β-diketones undergo monoprotonation. Using the acid system $HF-SbF_3$, the β-diketones H(hfa), H(tfa), and H(tta) undergo conversion to the allylic carbonium ion (2) [28]. Similar results have been reported using the system $SO_2-SbF_5-FSO(OH)_2$ [29].

R = CH_3, CF_3, 2-thienyl

(2)

Polarographic reduction of H(tta) has been studied under a variety of conditions [30-32]. Four distinct waves have been determined in the pH range 2-10 [32]. The first wave is due to a one-electron reduction of the carbonyl group adjacent to the $-CF_3$ group in acid or neutral solution and also to reduction adjacent to the thienyl group at pH above 7. The second polarographic wave represents a one-electron reduction of the free radical at pH 3-4. The third and fourth waves are due to diffusion-controlled processes in which the following three species are reduced: the enolate form of tta$^-$ and the hydrated anions thienyl-$CO-CH_2C(O^-)(O^-)CF_3$ and thienyl-$COCH_2-C(O^-)(OH)CF_3$.

Spectrophotometry indicates that variation in the polarographic behavior of H(tta) as a function of alkalinity probably is due to a catalytic effect upon the pertinent rate constants for the conversion of ionic nonreducible species to un-ionized reducible intermediates [31].

C. Synthetic Applications Not Involving Metals

Fluorinated-β-diketones have been employed as synthetic reagents in a variety of transformations.

1. Xanthydrol [(3a), X=O] and thioxanthydrol [(3b), X=S] react with benzoyl-trifluoroacetone [H(bta)] to furnish the condensation products (4a, 4b) with elimination of H_2O [33].

(3a, 3b)   (4a, 4b)

2. Condensation of 2-aceto-p-cresol, 2-hydroxy-4, 5-dimethylacetophenone, and 2-hydroxy-4-methoxyacetophenone with ethyl trifluoroacetate in the presence of powdered sodium leads to production of the corresponding dike-tones (5), which undergo cyclization in refluxing ethanol containing HCl to yield the corresponding chromones (6) [34].

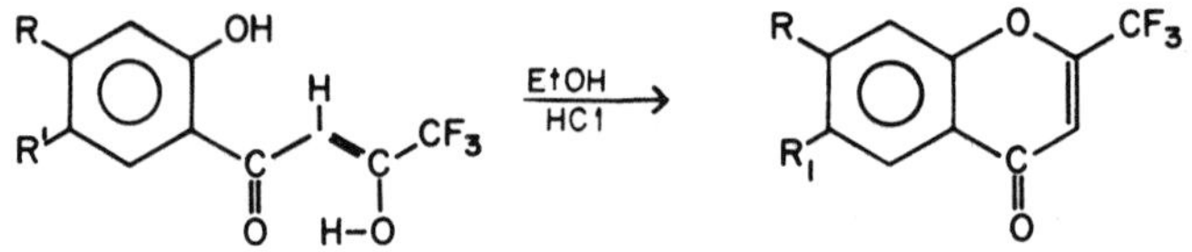

R = H; R' = CH$_3$
R = R' = CH$_3$
R = -O-CH$_3$; R' = H
(6)

Treatment of (6) with hydroxylamine yields the oximes (7) [34].

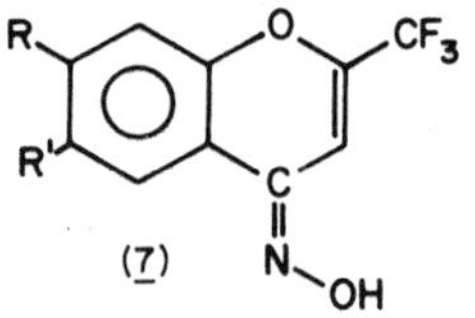

(7)

3. Reaction of a fluoro-β-diketone with hydrazine hydrate in an acid alcoholic medium results in formation of various pyrrazoles (8) [35, 36].

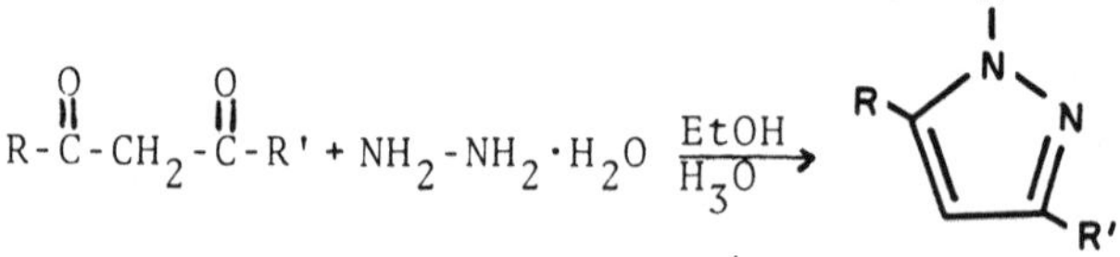

R=CF$_3$; a) R'=phenyl,   b) 3-benzothienyl
        c) 2,3 or 4-pyridyl

(8)

The action of peracetic acid on (8b) gives the dioxo benzothienyl derivatives [35, 36] which have been found to be effective agents against bacteria, fungi, and algae.

4. The fluorinated diaryl β-diketone (9) is converted by a series of transformations to the cyclic fluorosilicon product (10) [37].

5. Treatment of diketone (11) with a mixture of 100% HNO_3 and conc H_2SO_4
at low temperature gives the ring-nitrated derivatives (12) [38].

6. Furyl and thienyl derivatives of 4-trifluoromethyl-2-amino-pyrimidine
(13) are obtained when various fluoro-β-diketones are permitted to react with
guanidine carbonate in the presence of sodium methoxide [39].

R = 2-furyl, R' = H, CH_3
R = 2-thienyl, R' = H, $n-C_3F_7$

(13)

7. p-Fluorobenzene azodiketones (14) are condensed with hydrazine hydrate
to give the pyrazoles (15) and with hydroxylamine the isoxazole (16) [40].

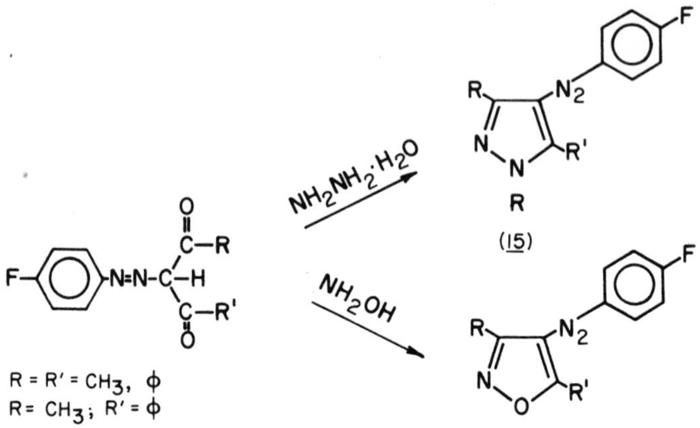

(15)

(16)

(14)

R = R' = CH$_3$, ϕ
R = CH$_3$; R' = ϕ

8. 2-Pyrroyltrifluoroacetone (17) reacts in the presence of hydroxylamine to yield the ixoxazoles (18) [41].

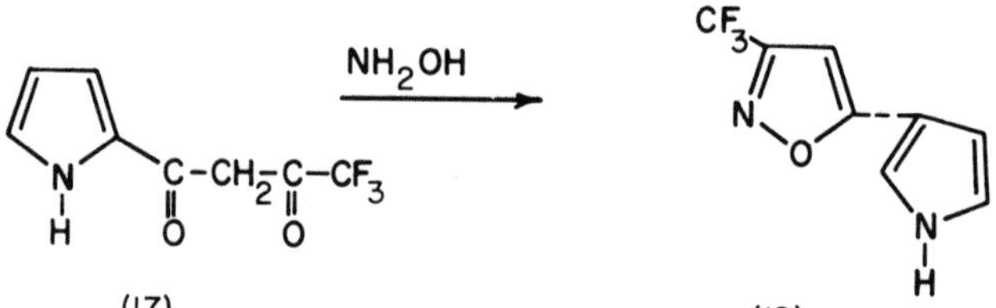

(17)

(18)

9. Heating the vinyl diketone (19) with a cross-linking monomer, divinyl benzene, gives a polymeric material [42].

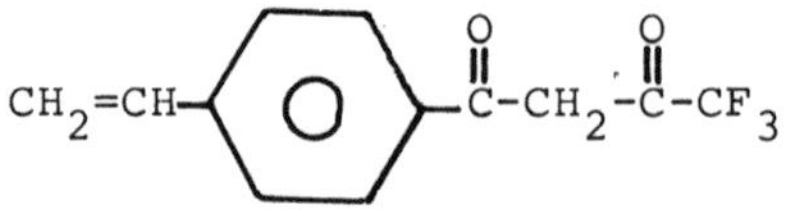

(19)

D. Physical Properties of Fluoro-β-Diketones

Physical properties of fluoronated β-diketones have been compiled in Table 2.

TABLE 2

Physical Properties of Selected Fluoro-β-Diketones

Fluoro-β-diketone	Density [Ref.] (g/ml)	Boiling point (mp) (°C)	Ref.
Fluoroacetylacetone		25 (15 mm)	[43]
Trifluoroacetylacetone [H(tfa)]	1.271 (23°C) [11]	107 (760 mm)	[12]
Hexafluoroacetylacetone [H(hfa)]	1.480 (23°C) [22]	70.0–70.2 (760 mm)	[22]
Hexafluoroacetylacetone dihydrate		<115 (subl.)	[23]
Octafluorohexanedione [H(ofhd)]	1.538 (23°C) [44]	85 (760 mm)	[44]
Octafluorohexanedione dihydrate		60–61 (mp)	[44]
Decafluoroheptanedione [H(dhfd)]	1.592 (23°C) [44]	99–105	[45]
Heptafluorodimethyloctanedione [H(fod)]	1.292 (23°C) [44]	33 (2.7 mm)	[46]
		56.5–64 (12–14 mm)	[44]
		43–49 (2–4 mm)	[47]
1,1,1-Trifluoro-3-methyl-2,4-pentane-2,4-dione	1.220 [12]	123 (755 mm)	[12]
1,1,1-Trifluoro-2,4-hexanedione	1.220 [12]	124 (755 mm)	[12]
1,1,1-Trifluoro-3-methyl-2,4-hexanedione	1.220 [12]	74 (43 mm)	[12]
1,1,1-Trifluoro-6-methyl-2,4-heptanedione	1.130 [12]	78 (64 mm)	[12]

TABLE 2 (Continued)

Physical Properties of Selected Fluoro-β-Diketones

Fluoro-β-diketone	Density [Ref.] (g/ml)	Boiling point (mp) (°C)	Ref.
1,1,1-Trifluoro-2,4-decanedione	1.090 [12]	65 (6 mm)	[12]
1,1,1-Trifluoro-4-phenyl-2,4-butanedione [H(bta)]		224 (760 mm)	[12]
		39-40.5 (mp)	[12]
2-Thenoyltrifluoroacetone		42.5-43.2 (mp)	[12]
1,1,1,2,2,3,3-Heptafluoro-4,6-heptanedione	1.3646 [7]	39-42 (15-16 mm)	[48]
		55.5-56.8 (38-39 mm)	[7]
1,1,1-Trifluoro-5-methyl-2,4-hexanedione		28-30 (1-2 mm)	[48]
1,1,1-Trifluoro-5,5-dimethyl-2,4-hexanedione		138-141 (760 mm)	[49]
		36-37 (1-2 mm)	[48]
1,1,1,2,2-Pentafluoro-6-methyl-3,5-heptanedione		38-39 (12 mm)	[48]
1,1,1,2,2-Pentafluoro-6,6-dimethyl-3,5-heptanedione		49 (6 mm)	[48]

Compound	Property	Ref.
1,1,1,5,5,6,6,7,7,7-Decafluoro-2,4-heptanedione	105 (760 mm)	[48]
1,1,1,2,2,3,3-Heptafluoro-4,6-octanedione	51–53 (15–16 mm)	[48]
1,1,1,2,2,3,3-Heptafluoro-7-methyl-4,6-octanedione	44–45 (6–7 mm)	[48]
1,1,1,2,2,6,6,7,6,8,8-Docafluoro-3,5-octanedione	52–58 (3–4 mm)	[48]
2-Trifluoroacetylindanone-1	72.5–73 (mp)	[7]
2-Trifluoroacetyltetralone-1	51–52 (mp)	[7]
2-Pyrroyltrifluoroacetone	68.5–70.0 (mp)	[41]
p-Fluorobenzoyltrifluoroacetone	40–42 (mp)	[12]
1,1,1-Trifluoro-4(p-biphenylyl)-2,4-butanedione	100.1–101.1 (mp)	[12]
2-Naphthoyltrifluoroacetone	70.1–71.1 (mp)	[12]
2-Furoyltrifluoroacetone	19–21 (mp)	[12]
4,4,4-Trifluoro-1-(4-methoxyphenyl)-1,3-butanedione	57–58 (mp)	[50]
4,4,4-Trifluoro-1-(2-methoxyphenyl)-1,3-butanedione	44–45 (mp)	[50]
4,4,4-Trifluoro-1-(2,5-dimethoxyphenyl)-1,3-butanedione	69–69.5 (mp)	[50]
4,4,4-Trifluoro-1-(2-fluorenyl)-1,3-butanedione	109–109.5 (mp)	[50]
4,4,4-Trifluoro-1-(3-phenanthryl)-1,3-butanedione	116–117 (mp)	[50]
1,1,1,2,2,3,3,4,4-Nonafluoro-5,7-octanedione	68–68.5 (33–34 mm) 1.3589 (23°C) [7]	[50]

TABLE 2 (Continued)

Physical Properties of Selected Fluoro–β–Diketones

Fluoro–β–diketones	Density [Ref.] (g/ml)	Boiling point (mp) (°C)	Ref.
1,1,1,2,2-Pentafluoro-3,5-hexanedione	1.3729 (23°C) [7]	111–112 (631 mm)	[7]
2-Trifluoroacetylcyclopentanone	1.4312 (23°C) [7]	66 (21–22 mm)	[7]
2-Trifluoroacetylcyclohexanone	1.4522 (23°C) [7]	76–77 (20–21 mm)	[7]
Thio Derivatives of Fluoro–β–diketones			
1,1,1-Trifluoro-4-(2-thienyl)-4-mercaptobut-3-en-2-one		62–64 (mp)	[5]
1,1,1-Trifluoro-4-mercaptopent-3-en-2-one		74 (mp)	[1]
1,1,1-Trifluoro-4-mercapto-4-phenylbut-3-en-2-one		50 (20 mm)	[2]
1,1,1-Trifluoro-4-mercapto-4-((4-methylphenyl)-but-3-en-2-one		95 (5 mm)	[3]
1,1,1-Trifluoro-4-mercapto-4-((4-methoxyphenyl)-but-3-en-2-one		Dec.	[4]
4-(4-Bromophenyl)-1,1,1-trifluoro-4-mercaptobut-3-en-2-one		Dec.	[4]
1,1,1-Trifluoro-4-(2-furyl)-4-mercaptobut-3-en-2-one		110 (2 mm)	[4]
		102 (1 mm)	[4]

E. Miscellaneous Properties of Fluoro-β-Diketones

Biochemically, the fat-soluble chelating agent H(tta) has a pronounced inhibitory effect on succinic oxidase activity. This β-diketone also promotes ATPase activity and uncouples phosphorylation from oxidation [51-53]. Reversal of inhibition of succinic oxidase can be effected using a large stoichiometric amount of coenzyme Q [52]. Trifluoroacetylacetone also demonstrates this inhibitory effect, though not as effectively as that of H(tta) [51]. The inhibitory effect is envisioned as being due to the lipophilic nature of these β-diketones and the chelating ability of the compounds for iron(II).

Several of the title reagents, notably 2-furoyltrifluoroacetone [H(fta)] have been observed to retard neoplasm formation in the Ehrlich Ascites tumor [54].

III. FLUORO-β-DIKETONES
AS METAL CHELATING AGENTS

A. Synthesis of the Metal Chelates

Three general methods have been reported for the preparation of metal fluoro-β-diketonates. Differences in reaction conditions exist from metal to metal within each general method, and such differences include choice of solvent, pH, and buffer, as well as methods for isolating the product.

1. Reaction of a Metal Salt with the Fluoro-β-Diketonates

This method was first used by Staniforth and co-workers [10, 55] for the preparation of the fluoro-β-diketonates of beryllium(II), aluminum(III), strontium(III), yttrium(II), lanthanum(III), cerium(III), prometheum(III), neodymium(III), samarium(III), europium(III), gadolinium(III), uranium(IV), and copper(II). The chelate is precipitated from a concentrated metal salt solution by the addition of the fluoro-β-diketone. The metal salt employed is usually the nitrate, oxalate, oxychloride, or acetate and the reaction is carried out in either aqueous, aqueous methanolic, or aqueous ethanolic solution. Recrystallization of the precipitate produces high yields of pure product.

2. Reaction of Anhydrous Metal Chlorides with the Fluoro-β-Diketone

This procedure, which was first described by Sievers and co-workers [56, 57], involves adding the fluoro-β-diketone directly to an anhydrous suspension of a metal chloride in carbon tetrachloride. The reaction mixture usually is refluxed and hydrogen chloride is evolved.

$$ZrCl_4 + 4CF_3\text{-}\overset{O}{\overset{\|}{C}}\text{-}CH_2\text{-}\overset{O}{\overset{\|}{C}}\text{-}CH_3 \xrightarrow[\text{reflux}]{CCl_4} Zr(CF_3\text{-}\overset{O}{\overset{\|}{C}}=CH\text{-}\overset{O}{\overset{\|}{C}}\text{-}CH_3)_4 + 4HCl$$

3. Reactions of Metal Carbonyls with Fluoro-β-Diketones

Dunne and Cotton [58] used this method for preparing molybdenum and chromium trifluoroacetylacetonates. The metal hexacarbonyl is refluxed for 2-4 days with H(tfa). Unreacted β-diketone is removed via vacuum distillation and the residual metal β-diketonate is recrystallized. The reaction apparently has the following stoichiometry:

$$M(CO)_6 + 3CF_3\text{-}\overset{O}{\overset{\|}{C}}\text{-}CH_2\text{-}\overset{O}{\overset{\|}{C}}\text{-}CH_3 \longrightarrow M(CF_3\text{-}\overset{O}{\overset{\|}{C}}=CH\text{-}\overset{O}{\overset{\|}{C}}\text{-}CH_3)_3 + 3/2H_2$$
$$+ 6CO$$

$$M = Co, Mo$$

4. Special Methods for Mixed Metal Chelates

The chemical diversity of the compounds precludes a detailed description of individual compounds as there are no general methods of preparation. Two examples of methods employed are given below and a tabulation of many mixed metal chelates together with their physical properties is given in Tables 3 and 4.

Melby and Rose [59] prepared mixed eight coordinated trivalent rare-earth β-diketone metal chelates by reacting the β-diketone chelate with other ligands such as pyridinium, triethylamine, and 2,4,6-trimethylpyridinium.

Boucher and Bailar [60] prepared 1,1,1-trifluoro-2,4-pentanedionatobis (ethylenediamine)cobalt(III) iodide by treating cis-chloroaquobis(ethylenediamine)cobalt(III)bromide with H(tfa) at 40°C, followed by the addition of potassium iodide to precipitate the iodide.

B. Physical Properties of Fluoro-β-Diketonates

Some of the more important physical properties of metal fluoro-β-diketonates are listed in Tables 5, 6, and 7.

C. Chemical Spectral Studies
of Metal Fluoro-β-Diketonates

The range of spectral studies which have been carried out with a wide variety of metals are described in this section. Emphasis is placed on those investigations which have yielded data of value in structural or electronic transition assignments. Not included in this survey are ESR [96] and ORD-CD [97] spectral studies which have been recently covered in detail. Fluorescence and related spectral investigations are included due to the current interest in the use of rare earth fluoro-β-diketonates in laser technology.

TABLE 3

Physical Properties of Selected Mixed Chelates

Structure	R_1	R_2	R_3	R_4	Metal	Melting point (°C)	Ref.
	CF_3	H	CH_3	CH_3	Cu(II)	196–197 dec.	[84]
	CF_3	H	CH_3	$n-C_4H_9$	Cu(II)	180–181 dec.	[84]
	CF_3	H	OC_2H_5	$n-C_4H_9$	Cu(II)	171–173 dec.	[84]
	CF_3	H	CF_3	$n-C_4H_9$	Cu(II)	135–140 dec.	[84]
	CF_3	H	CH_3	$i-C_4H_9$	Cu(II)	168–170 dec.	[84]
	CF_3	H	CH_3	$s-C_4H_9$	Cu(II)	180–181 dec.	[84]
	CH_3	H	CF_3		Ir(II) (green)	148 (60 subl.)	[94]
	CF_3	H	CF_3		Ir(II) (green)	92 (40 subl.)	[94]
	C_4H_4S	H	CF_3		Ir(II) (green)	160 dec. (100 subl.)	[94]
	2-Thienyl	H	CF_3	$(C_2H_5)_3NH$	Eu	133	[59]
	2-Thienyl	H	CF_3	2,4,6,-Collidinium	Eu	158	[59]
	2-Thienyl	H	CF_3	$(C_2H_5)_3NH$	Tb	158	[59]
	2-Thienyl	H	CF_3	Pyridinium	Tb	193	[59]

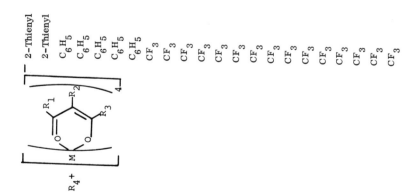

$$R_4^+ \left[M \left(R_1,\ R_2,\ R_3 \right) \right]_4$$

R_1	R_2	R_3	R_4^+	M		Ref.
2-Thienyl	H	CF$_3$	$(C_2H_5)_3NH$	Nd	135	[59]
2-Thienyl	H	CF$_3$	$(C_2H_5)_3NH$	La	135	[59]
C_6H_5	H	CF$_3$	$(C_2H_5)_3NH$	Eu	108	[59]
C_6H_5	H	CF$_3$	$(C_2H_5)_4N$	Eu	152	[59]
C_6H_5	H	CF$_3$	$(C_2H_5)_3NH$	Tb	130	[59]
C_6H_5	H	CF$_3$	$(C_2H_5)_3NH$	La	130	[59]
C_6H_5	H	CF$_3$	$(C_2H_5)_3NH$	Nd	130	[59]
CF$_3$	H	CF$_3$	$(C_2H_5)_3NH$	Eu	130	[59]
CF$_3$	H	CF$_3$	Pyridinium	Eu	175	[59]
CF$_3$	H	CF$_3$	2,6-Lutidinium	Eu	130	[59]
CF$_3$	H	CF$_3$	2,4,6-Collidinium	Eu	130	[59]
CF$_3$	H	CF$_3$	Piperazinium	Eu	210	[59]
CF$_3$	H	CF$_3$	$(C_2H_5)_4N$	Eu	153	[59]
CF$_3$	H	CF$_3$	$(CH_3)_4N$	Eu	190	[59]
CF$_3$	H	CF$_3$	N-methylquinolinium	Eu	100	[59]
CF$_3$	H	CF$_3$	N-methylphenazinium	Eu	120	[59]
CF$_3$	H	CF$_3$	$(C_2H_5)_3NH$	Tb	130	[59]
CF$_3$	H	CF$_3$	Pyridinium	Tb	185	[59]
CF$_3$	H	CF$_3$	2,6-Lutidinium	Tb	130	[59]
CF$_3$	H	CF$_3$	2,4,6-Collidinium	Tb	135	[59]
CF$_3$	H	CF$_3$	$(C_2H_5)_3NH$	Nd	130	[59]
CF$_3$	H	CF$_3$	Pyridinium	Nd	185	[59]

TABLE 3 (Continued)

Structure	R_1	R_2	R_3	R_4	Metal	Melting point (°C)	Ref.
	CF_3	H	CF_3	2,6-Lutidinium	Nd	130	[59]
	CF_3	H	CF_3	$(C_2H_5)_3NH$	La	130	[59]
	CF_3	H	CF_3	2,6-Lutidinium	La	130	[59]
	CF_3	H	CF_3	$(C_2H_5)_3NH$	Pr	130	[59]
	CF_3	H	CF_3	$(C_2H_5)_3NH$	Sm	130	[59]
	CF_3	H	CF_3	$(C_2H_5)_3NH$	Gd	130	[59]
	CF_3	H	CF_3	$(C_2H_5)_3NH$	Dy	130	[59]
	CF_3	H	CF_3	$(C_2H_5)_3NH$	Ho	130	[59]
	CF_3	H	CF_3	$(C_2H_5)_3NH$	Er	130	[59]
	CF_3	H	CF_3	$(C_2H_5)_3NH$	Yb	130	[59]
	2-Thienyl	H	CF_3	2,4,6-Trimethylpyridinium	Eu	159–160	[89]
	2-Thienyl	H	CF_3	2,4,6-Trimethylpyridinium	Tb	155–157	[89]
	2-Thienyl	H	CF_3	Isoquinolinium	Eu	170–171	[89]
	2-Thienyl	H	CF_3	$(n\text{-}C_3H_7)_4N$	Eu	188–189	[89]
	2-Thienyl	H	CF_3	$(n\text{-}C_6H_{13})_4N$	Eu	170–172	[89]
	C_6H_5	H	CF_3	2,4,6-Trimethylpyridinium	Eu	168–170	[89]
	C_6H_5	H	CF_3	Isoquinolinium	Eu	150–151	[89]
	$-C_{10}H_7$	H	CF_3	Isoquinolinium	Eu	192–196	[89]
	CH_3	H	CF_3	Isoquinolinium	Eu	111–114	[89]
	C_6H_5	H	CF_3	**Piperidinium**	Eu	170–172	[90]

C_6H_5	H	CF_3	n-Butylammonium	Eu	78-80	[90]
C_6H_5	H	CF_3	Diethylammonium	Eu	135-136	[90]
$C_6H_5 \cdot H_2O$	H	CF_3	Triethylammonium	Eu	87-88	[90]
C_6H_5	H	CF_3	Benzylammonium	Eu	68-70	[90]
C_6H_5	H	CF_3	Dibenzylammonium	Eu	152-153	[90]
C_6H_5	H	CF_3	Tetramethylammonium	Eu	217-218	[90]
C_6H_5	H	CF_3	Tetra-n-propylammonium	Eu	167-168	[90]
C_6H_5	H	CF_3	Tetra-n-butylammonium	Eu	136-137	[90]
C_6H_5	H	CF_3	2-Hydroxyethylammonium	Eu	171-172	[90]
C_6H_5	H	CF_3	Tetramethylguanidinium	Eu	154-155	[90]
C_6H_5	H	CF_3	Pyridinium	Eu	180-185	[90]
C_6H_5	H	CF_3	Quinolinium	Eu	155-157	[90]
4-Fluorophenyl	H	CF_3	Piperidinium	Eu	157-160	[50]
4-Methoxyphenyl	H	CF_3	Piperidinium	Eu	217-219	[50]
2-Methoxyphenyl	H	CF_3	Piperidinium	Eu	117-120	[50]
5,5-Dimethoxyphenyl	H	CF_3	Piperidinium	Eu	180-182	[50]
4-Biphenyl	H	CF_3	Piperidinium	Eu	237-238	[50]
1-Naphthyl	H	CF_3	Piperidinium	Eu	170-175	[50]
2-Naphthyl	H	CF_3	Piperidinium	Eu	203-205	[50]
2-Fluorenyl	H	CF_3	Piperidinium	Eu	205-208	[50]
3-Phenanthryl $\cdot H_2O$	H	CF_3	Piperidinium	Eu	145-149	[50]
2-Thienyl	H	CF_3	Piperidinium	Eu	164-165	[50]

TABLE 4

Physical Properties of Miscellaneous Mixed Chelates

Chelate[a]	Melting point (°C)	Color	Ref.
$Eu(tta)_3(Phen)$	247–249		[59]
$Eu(tta)_3(Dipy)$	221–224		[59]
$Eu(tta)_3(Tripy)$	247–251		[59]
$Eu(tta)_3(PicNO)_2$	234–236		[59]
$Eu(tta)_3(TPPO)_2$	251–253		[59]
$Eu(tta)_2(TPPO)_2NO_3$			[59]
$Tb(tta)_2(TPPO)_2NO_3$	232–234		[59]
$Cu(tfa)_2(4Me-py)_2$	130	Green	[85]
$Cu(hfa)_2(4Me-py)_2$	145	Green	[85]
$Cu(hfa)_2(py)_2$		Green	[85]
$Cu(hfa)_2 \cdot 4Me-py$	55	Green	[85]
$Cu(tfa)_2(py)_2$			[86]

Cu(tfa)₂·py			[86]
Cu(tfa)₂(Quin)₂			[86]
Cu(tfa)₂·acet			[86]
Mn(CO)₂[P(C₆H₅)₃]₂(hfa)	159	Red–brown	[87]
Mn(CO)₂[P(C₆H₅)₂(CH₃)]₂(hfa)	120	Red–brown	[87]
Mn(CO)₂[P(C₆H₅)(CH₃)₂]₂(hfa)	115	Red–brown	[87]
Mn(CO)₂[P(n-C₄H₉)₃]₂(hfa)	72	Red–brown	[87]
Mn(CO)₂[P(OCH₃)₃]₂(hfa)	40	Purple	[87]
Mn(CO)₂[P(OC₆H₅)₃]₂(hfa)	90	Orange	[87]
Mn(CO)₂[P(O-n-C₄H₉)₃]₂(hfa)		Orange	[87]
Mn(CO)₃[P(C₆H₁₁)₃](hfa)	128	Red–brown	[87]
Mn(CO)₃[As(C₆H₅)₃](hfa)	64	Orange	[87]
Mn(CO)₃(C₆H₅N)(hfa)	51	Orange	[87]
Mn(CO)₃(4–CH₃C₅H₄N)(hfa)	75	Orange	[87]
Zn(bta)₂TPPO	168		[63]
Cd(bta)₂TPPO	185		[63]
Cd(bta)₂TOPO	Oil		[63]
Cd(hfa)₃·NH₄·H₂O	178-179	White	[95]

TABLE 4 (Continued)

Physical Properties of Miscellaneous Mixed Chelates

Chelate	Melting point (°C)	Color	Ref.
$Cd(hfa)_2 \cdot NH_3 \cdot H_2O$	168–171	White	[95]
$Zn(bta)_2 TOPO$	Oil		[63]
$Fe(OCH_3)_2 CF_3 COCHCOC_4H_3S$		Yellow	[91]
$ReCl_2(tfa)(PPh_3)_2$	191–192	Red–purple	[92]
$ReCl_2(tta)(PPh_3)_2$	188–192	Lilac	[92]
$ReCl_2(hfa)(PPh_3)_2$	124–127	Blue	[92]
Bis(2, 4–pentanedionato)(3–fluoro–2, 4–pentanedionato)chromium(III)	212.5–213.5		[93]
Trifluoroacetylacetonatobis(ethylenediamine)cobalt(III)iodide		Red–orange	[60]
Bis(trifluoroacetylacetonato)dimethylsulfoxidecopper(II)	111–113	Blue–green	[88]
Bis(trifluoroacetylacetonato)bis(dimethylsulfoxide)zinc(II)	94–98	White	[88]

[a] Abbreviations: Phen, 1, 10–phenanthroline; Dipy, dipyridyl; Tripy, tripyridyl; PicNO, 4–picoline N–oxide; TPPO, triphenylphosphine oxide; 4Me–py, 4–methyl–pyridine; py, pyridine; Quin, quinoline, acet, acetone; TOPO, tri–n–octylphosphine oxide; PPh_3, triphenylphosphine.

TABLE 5

Physical Properties of Selected Fluoro-β-Diketonates

Chelating agent	Metal	Color	Melting point (°C)	Sublimed (°C)	Ref.
H(tfa)	Cr(III), cis	Violet	112–114		[61]
			124–126		[58]
H(tfa)	Cr(III), trans	Violet	154.5–155		[61]
			150–151.5		[83]
H(tfa)	Co(III), cis	Green	129–129.5		[61]
H(tfa)	Co(III), trans	Green	158–158.5		[61]
H(tfa)	Rh(III), cis	Yellow	148.5–149		[61]
H(tfa)	Al(III), trans	White	121–122	48[79]	[61]
			117		[62]
H(tfa)	Ga(III), trans	White	128.5–129.5		[61]
H(tfa)	In(III), trans	Ivory	118–120		[61]
H(tfa)	Mn(III), trans	Green–brown	113–114		[61]
H(tfa)	Fe(III), trans	Red	114		[61]
			115	48[79]	[62]
H(tfa)	Be(II)	White	112	38[79]	[55]

TABLE 5 (Continued)

Physical Properties of Selected Fluoro-β-Diketonates

Chelating agent	Metal	Color	Melting point (°C)	Sublimed (°C)	Ref.
H(tfa)	Cu(II)	Blue-violet	189		[12]
	Ni(II)		200	55[79]	[62]
H(tfa)	Sc(III)	Green		150 (0.5 mm)	[62]
H(tfa)	Y(III)	White	106–107	90–95	[55]
H(tfa)	La(III)	White	132		[55]
H(tfa)	Ce(III)		169		[55]
H(tfa)	Pr(III)	Yellow	130–131		[55]
H(tfa)	Mo(III)	Yellow	133–134		[55]
H(tfa)	Nd(III)	Brown–black	155–157		[58]
H(tfa)	Sm(III)	Pink	133–134		[55]
H(tfa)		White			[55]

Ligand	Metal	Color	M.p. (°C)	B.p.	Ref.
H(tfa)	Eu(III)	White	132–134		[55]
H(tfa)	Gd(III)	White	133–135		[55]
H(tfa)	U(IV)	Olive	142–144		[55]
H(tfa)			138–140	115–120 (10^{-3} mm)	[49]
H(tfa)	Fe(II)	Red			[64]
H(tfa)	Zr(IV)	White	128–130	115 (0.05 mm)	[66]
H(tfa)	Hf(IV)	White	125–128	115 (0.05 mm)	[66]
H(tfa)	Pu(IV)				[68]
H(tfa)	Zn(II)	White	168–169		[75]
H(tfa)	Zn(II), dihydrate	White	186–187		[75]
H(tfa)	Co(II)·5/2H$_2$O	Red–brown			[76]
H(hfa)	Fe(II)	Red		45–50 (10^{-3} mm)	[64]
H(hfa)	Cu(II), monohydrate	Green	126–128		[65]
H(hfa)	Zn(II), dihydrate	White	153–154		[65]
H(hfa)	Ni(II), dihydrate		207–208		[65]
H(hfa)	Co(II), dihydrate		172–174		[65]
H(hfa)	Mn(II), dihydrate		155–156		[65]
H(hfa)	Fe(II), dihydrate		Dec.		[65]

TABLE 5 (Continued)

Physical Properties of Selected Fluoro-β-Diketonates

Chelating agent	Metal	Color	Melting point (°C)	Sublimed (°C)	Ref.
H(hfa)	Rh(III)	Yellow	114–115		[65]
			114.5–115		[83]
H(hfa)	Fe(III)	Red	49	35	[65]
H(hfa)	Cr(III)	Green	84–85	25	[65]
H(hfa)	Al(III)	White	73–74		[65]
H(hfa)	Nd(III)	Pink	117		[55]
H(hfa)	Nd(III), monohydrate		115–121		[65]
H(hfa)	Th(IV)		121–122		[65]
H(hfa)	Zr(IV)		152–154		[65]
H(hfa)	Be(II)		70–71		[67]
H(hfa)	La(III)	White	120–125		[55]
H(hfa)	Sm(III)	White	125		[55]

H(hfa)	U(IV)	Green-brown	90 (mp)	70–80 (0.2 mm) dec.	[43]
H(hfa)	Co(III)	Green	145 (bp) dec.	40–50 (10^{-3} mm)	[74]
H(hfa)	Cu(II)	Purple	94–95		[75]
H(hfa)	Cu(II), dihydrate	Green	95–98	113–115 (10)	[75]
H(bta)	Al(III)	White	134–136	88 (79)	[62]
H(bta)	Be(II)	White	173–174	78 (79)	[62]
H(bta)	Co(II)	Yellow	143–144	55 (99)	[62]
H(bta)	Cu(II)	Green	158	130 dec. (79)	[12]
H(bta)			243–244		[65]
H(bta)			241		[45]
H(bta)			237.5–238.5		
H(bta)	Fe(III)	Red	128–129	89 dec. (79)	[62]
H(bta)	Mn(II)	Yellow	129–130		[62]
H(bta)	Ni(II)	Green	223–224	98 (79)	[62]
H(bta)	Ni(II), dihydrate	Green			[62]
H(bta)	U(IV)	White	191 (bp) (0.003 mm)		[70]
H(fta)	Al(III)	White	204–205	95 (79)	[62]
H(fta)	Be(II)	White	169–170	71 (79)	[62]

TABLE 5 (Continued)

Physical Properties of Selected Fluoro-β-Diketonates

Chelating agent	Metal	Color	Melting point (°C)	Sublimed (°C)	Ref.
H(fta)	Co(II)	Yellow	215–220	63 (79)	[62]
H(fta)	Cu(II)	Green	226–228	105 (79)	[12]
			227–228		[62]
H(fta)	Fe(III)	Red	207–208	95 dec. (79)	[62]
H(fta)	Mn(II)	Yellow–orange	146–149		[62]
H(fta)	Ni(II)	Yellow	293–296	65 (79)	[62]
H(fta)	Ni(II), dihydrate	Green			[62]
H(fta)	Zr(IV)	White	199–201		[66]
H(fta)	Hf(IV)	White	195–197		[66]
H(tta)	Al(III)	White	203–205	125 (79)	[62]
H(tta)	Be(II)	White	169–170	73 dec. (79)	[62]
H(tta)	Co(II)	Yellow	200–215 dec.	69 (79)	[62]

H(tta)	Cu(II)	Green	242–243	112 dec. (79)	[62]
H(tta)	Fe(III)	Red	159–160	135 (79)	[62]
H(tta)	Mn(II)	Orange	177–179		[62]
H(tta)	Ni(II)	Yellow	291–295	70 (79)	[62]
H(tta)	Ni(II), dihydrate	Green			[62]
H(tta)	Zr(IV)	White	225–226		[66]
H(tta)			240		[77]
H(tta)	Hf(IV)	White	220–223		[66]
H(tta)			240		[77]
H(tta)	Eu(IV)	White	164–165		[50]
H(tta)	Th(IV)		224		[78]
H(tta)			235		[77]
H(tta)	U(IV)		250		[77]
H(tta)	Ti(III)	Green			[81]
H(tta)	La(III)	Brown	135		[73]
H(tta)	Ce(IV)	Brown	181		[73]
H(tta)			190		[77]
H(tta)	Pr(IV)	Brown	164		[73]

TABLE 5 (Continued)

Physical Properties of Selected Fluoro–β–Diketonates

Chelating agent	Metal	Color	Melting point (°C)	Sublimed (°C)	Ref.
H(tta)	Nd(III)	Pink	181–182		[73]
H(tta)	Sm(III)	Straw	148		[73]
H(tta)	Eu(III)	Pink	180		[73]
H(tta)	Gd(III)	Grey	122		[73]
H(tta)	Tb(III)	Pink	115		[73]
H(tta)	Dy(III)	Yellow	193		[73]
H(tta)	Ho(III)	Brown	135		[73]
H(tta)	Er(III)	Pink	125		[73]
H(tta)	Tm(III)	Brown	115		[73]
H(tta)	Yb(III)	Yellow	145		[73]
H(fod)	Sc(III)		25		[46]
	Sc(III), monohydrate		25		

H(fod)	Lu(III)	118–125	[46]
H(fod)	Lu(III), monohydrate	111–115	
H(fod)	Yb(III)	125–132	[46]
H(fod)	Yb(III), monohydrate	112–115	[46]
H(fod)	Tm(III)	140–146	[46]
H(fod)	Tm(III), monohydrate	110–115	
H(fod)	Er(III)	158–164	[46]
H(fod)	Er(III), monohydrate	104–112	
H(fod)	Y(III)	162–167	[46]
H(fod)	Y(III), monohydrate	108–112	[46]
H(fod)	Ho(III)	172–178	[46]
H(fod)	Ho(III), monohydrate	103–111	
H(fod)	Dy(III)	180–188	[46]
H(fod)	Dy(III), monohydrate	103–107	
H(fod)	Tb(III)	190–196	[46]
H(fod)	Tb(III), monohydrate	92–97	

TABLE 5 (Continued)

Physical Properties of Selected Fluoro-β-Diketonates

Chelating agent	Metal	Color	Melting point (°C)	Sublimed (°C)	Ref.
H(fod)	Gd(III)		203-213 dec.		[46]
H(fod)	Gd(III), monohydrate		60-65		[46]
H(fod)	Eu(III)		205-212 dec.		[46]
H(fod)	Eu(III), monohydrate		59-67		[46]
H(fod)	Sm(III)		208-218 dec.		[46]
H(fod)	Sm(III), monohydrate		63-67		[46]
H(fod)	Nd(III)		210-215 dec.		[46]
H(fod)	Pr(III)		218-225 dec.		[46]
H(fod)	Pr(III), monohydrate		105-125		[46]
H(fod)	La(III)		215-230 dec.		[46]
H(fod)	U(IV)	Olive-brown	148-150		[71]
H(fod)	Th(IV)	White	147-148		[71]
H(dhfd)	Cu(II)		73-83	50 (0.1 mm)	[45]
H(dhfd)	Ca(II)			100 (0.005 mm)	[45]
H(dhfd)	Sr(II)		170-186		[45]

H(dhfd)	Ba(II)		145–180 dec.		[45]
H(dhfd)	Mg(II)		175–192	135 (0.005 mm)	[45]
H(dhfd)	Zr(IV)	Liquid	65–70 (bp) (0.005 mm)		[45]
H(dhfd)	Hf(IV)	Liquid	70 (bp) (0.02 mm)		[45]

TABLE 6

Physical Properties of Miscellaneous Fluoro–β–Diketonates

Chelate	Color	Melting point (°C)	Sublimed (°C)	Ref.
Cu(II) of 5, 5, 5–trifluoro–3–methyl–2, 4–pentanedione	Green	170.4–171.9		[12]
Cu(II) of 6, 6, 6–trifluoro–3, 5–hexanedione	Blue	154.5–155.3		[12]
Cu(II) of 6, 6, 6–trifluoro–4–methyl–3, 5–hexanedione	Green	164–165		[12]
Cu(II) of (i–C$_4$H$_9$)trifluoroacetone	Blue	124.5–125.3		[12]
Cu(II) of (n–C$_6$H$_{13}$)trifluoroacetone	Blue	71–72		[12]
Cu(II) of p–xenyltrifluroracetone	Green	303		[12]
Cu(II) of p–fluorobenzoyltrifluoroacetone	Green	263–264		[12]
Cu(II) of 2–naphthoyltrifluoroacetone	Green	278.5–279.5		[12]
Cu(II) of 2–thenoylperfluorobutyrylmethane	Green	Dec.		[69]
Cu(II) of 1, 1, 1, 2, 2, 6, 6, 6–octafluoro–3, 5–hexanedione	Green	76–79	25 (0.005 mm)	[45]
Cu(II) of 1, 1, 1, 2, 2, 2–pentafluoro–3, 5–hexanedione		111.6–112		[7]
Cu(II) of 1, 1, 1, 2, 2, 3, 3–heptafluoro–4, 6–heptanedione		55–55.2		[7]

Compound	Appearance	mp	bp	Ref.
Cu(II) of 1,1,1,2,2,3,3,4,4-nonafluoro-5,7-octanedione		77.5-78		[7]
Cu(II) of 2-trifluoroacetylcyclopentanone		175-176		[7]
Cu(II) of 2-trifluoroacetylcyclohexanone		182-182.5		[7]
Cu(II) of 2-trifluoroacetylindanone-1		272		[7]
Cu(II) of 2-trifluoroacetyltetralone-1		291-292		[7]
Cu(II) of 1,1,1-trifluoro-5,5-dimethyl-2,4-hexanedione		106	90 (0.1 mm)	[72]
Cr(III) of 1,1,1,2,2,6,6,6-octafluoro-3,5-hexanedione	Liquid		25 (bp) (0.001 mm)	[45]
Zr(IV) of 2-pyrroyltrifluoroacetone	White	184-185		[66]
Hf(IV) of 2-pyrroyltrifluoroacetone	White	185-186		[66]
U(IV) of 1,1,1-trifluoro-5,5-dimethyl-2,4-hexanedione		138		[49]
U(IV) of 1,1,1-trifluoro-2,4-hexanedione		60	145 (bp) (0.002 mm)	[70]
U(IV) of 1,1,1-trifluoro-2,4-heptanedione		15	116 (bp) (0.001 mm)	[70]
U(IV) of 1,1,1-trifluoro-5-methyl-2,4-hexanedione		78	132 (bp) (0.001 mm)	[70]
U(IV) of 1,1,1-trifluoro-2,4-octanedione			134 (bp) (0.003 mm)	[70]
			142 (bp) (0.005 mm)	[70]

TABLE 6 (Continued)

Physical Properties of Miscellaneous Fluoro-β-Diketonates

Chelate	Color	Melting point (°C)	Sublimed (°C)	Ref.
U(IV) of 1,1,1-trifluoro-6-methyl-2,4-heptanedione		82		[70]
		141 (bp) (0.002 mm)		
U(IV) of 1,1,1-trifluoro-2,4-nonanedione		166 (bp) (0.004 mm)		[70]

TABLE 7

Physical Properties of Selected Monothio Analogs

Chelate	Color	Melting point (°C)	Sublimed (°C)	Ref.
Co(III) of thiotrifluoroacetylacetone	Dark brown	131		[2]
Ni(II) of thiotrifluoroacetylacetone	Brown	153		[2]
Pd(II) of thiotrifluoroacetylacetone	Orange	154		[2]
Pt(II) of thiotrifluoroacetylacetone	Red	144		[2]
Cu(I) of thiotrifluoroacetylacetone	Red–brown	202		[2]
Zn(II) of thiotrifluoroacetylacetone	Yellow	101		[2]
Cd(II) of thiotrifluoroacetylacetone	Yellow	197		[2]
Hg(II) of thiotrifluoroacetylacetone	Yellow	146 dec.		[2]
Fe(III) of thiothenoyltrifluoroacetone	Black	165		[2]
Co(III) of thiothenoyltrifluoroacetone	Dark brown	235		[2]
Pt(II) of thiothenoyltrifluoroacetone	Brown	218		[2]
Cd(II) of thiothenoyltrifluoroacetone	Yellow–brown	213 dec.		[2]
Hg(II) of thiothenoyltrifluoroacetone	Yellow	178		[2]
Pb(II) of thiothenoyltrifluoroacetone	Yellow–brown	154 dec.		[2]

Table 7 (Continued)

Physical Properties of Selected Monothio Analogs

Chelate	Color	Melting point (°C)	Sublimed (°C)	Ref.
Cu(II) of thiothenoyltrifluoroacetone	Olive-brown	230–231		[5]
		187		[2]
Ni(II) of thiothenoyltrifluoroacetone	Brown-red	235–237		[5]
Zn(II) of thiothenoyltrifluoroacetone	Yellow-green	177–178		[5]
		183		[2]
Co(II) of thiothenoyltrifluoroacetone	Brown-blue	229		[5]
Pd(II) of thiothenoyltrifluoroacetone	Brown-red	236		[5]
		246		[2]
Ni(II) of (p-methylphenyl)-thiothenoyltrifluoroacetone	Brown	200		[4]
Ni(II) of (p-methoxyphenyl)-thiothenoyltrifluoroacetone	Brown	161		[4]
Ni(II) of (p-bromophenyl)-thiothenoyltrifluoroacetone	Dark brown	221		[4]

Compound	Color		Ref
Ni(II) of (2-furyl)-thiothenoyltrifluoroacetone	Brown	216	[4]
Cu(I) of (p-methylphenyl)-thiothenoyltrifluoroacetone	Brown	175	[4]
Cu(I) of (p-methoxyphenyl)-thiothenoyltrifluoroacetone	Dark brown	181	[4]
Cu(I) of (p-bromophenyl)-thiothenoyltrifluoroacetone	Dark brown	178	[4]
Cu(II) of (p-methylphenyl)-thiothenoyltrifluoroacetone	Brown	162	[4]
Cu(II) of (p-methoxyphenyl)-thiothenoyltrifluoroacetone	Brown	170	[4]
Cu(II) of (p-bromophenyl)-thiothenoyltrifluoroacetone	Brown	156	[4]
Cu(II) of (2-furyl)-thiothenoyltrifluoroacetone	Black	157	[4]
Co(III) of (phenyl)-thiothenoyltrifluoroacetone	Black	168	[3]
Ni(II) of (phenyl)-thiothenoyltrifluoroacetone	Brown	164	[3]
Pd(II) of (phenyl)-thiothenoyltrifluoroacetone	Orange	190	[3]
Pt(II) of (phenyl)-thiothenoyltrifluoroacetone	Red	208	[3]
Cu(II) of (phenyl)-thiothenoyltrifluoroacetone	Brown-green	140	[3]
Zn(II) of (phenyl)-thiothenoyltrifluoroacetone	Yellow	140	[3]
Cd(II) of (phenyl)-thiothenoyltrifluoroacetone	Orange	154	[3]
Hg(II) of (phenyl)-thiothenoyltrifluoroacetone	Yellow	138	[3]
Ni(II) of thiohexafluoroacetylacetone		25 (760 mm)	[82]

1. Infrared Spectral Studies

In one of the first vibrational spectral studies of metal fluoro-β-diketonates, Bellamy and Branch [98] found that the carbonyl absorptions (two peaks at 1550-1600 cm^{-1} and 1280-1390 cm^{-1}) are not dependent on the double bond strength of the enol form of the β-diketone, and no direct frequency-stability relationship is observed.

Trifluoro- and hexafluoroacetylacetonates of divalent cobalt, copper, nickel, and palladium in the 420-480 cm^{-1} region display shifts to higher frequencies relative to the nonfluorinated derivatives, indicating that the band is due to one of the metal-oxygen stretching modes [99].

The infrared absorption of a number of copper(II) chelates derived from H(tfa), H(tta), H(hfa), H(fta), 2-thenoylperfluorobutyrylmethane, and 2-furoylperfluorobutyrylmethane have been studied. In bis(hexafluoroace-tylacetonato)copper(II) there are two strong bands at 1645 and 1615 cm^{-1}, possibly arising from perturbated carbonyl absorptions. The increase in frequency relative to the nonfluorinated chelate is due to the negative inductive effect of the trifluoromethyl groups. The 1608 and 1568 cm^{-1} bands of bis(2-furoyltrifluoroacetonato)copper(II) and the 1615 and 1585 cm^{-1} bands of bis(2-furoylperfluorobutyrylmethano)copper(II) may be explained by the electron-withdrawing effect of the fluoroalkyl groups combined with the aromatic substituent functioning through resonance as an electron sink [69]. Frequency shifts due to the trifluoromethyl group(s) have also been cited in a study of the uranyl(VI) chelates of H(tfa) and H(hfa) [100]. Nakamoto et al. [101], have studied the nickel(II) and copper(II) chelates of H(tfa) and H(hfa) in the 4000-300 cm^{-1} region and compared these values to those using a perturbation treatment. In addition to higher frequency shifts in the carbonyl stretching region, shifts to lower frequencies of the metal-oxygen band (400-500 cm^{-1}) are seen.

The hexafluoroacetylacetonates of rhodium(III), chromium(III), iron(III), aluminum(III), zirconium(IV), thorium(IV), nickel(II), neodymium(III), zinc(II), manganese(II), iron(II), zirconium(IV), and copper(II) display the most intense bands in the infrared region at 1600 cm^{-1}, which is assigned to the carbonyl

groups. For charge-to-radius (q/r) ratios> 3.6, bands occur in the 1625-

1650 cm^{-1} region; for q/r ratios< 3.6, in the 1590-1615 cm^{-1} region [65].

Infrared spectra studies of lanthanide complexes of the R_3M and $(R_4M)K$

type where R includes H(bta) and H(tta), M equals La, Tb, or Eu, and K

equals tetramethylammonium, triethylbenzylammonium, or piperidinium also

have been made [102]. From a comparison of tri- and tetra-ligand-containing

complexes, it is seen that all ligands are equivalent in relation to the metal.

Differences seen in the shifts of γ(C=O) and γ(C=C) for the complexes are

accounted for by the strength of the metal-ligand bond [102]. A similar study

carried out with samarium, gadolinium, and dysprosium showed spectra to

be identical, indicating little interaction among the chelating rings [103].

Spectra of the ternary complexes of H(bta) and tri-n-butyl or tri-n-octyl

phosphine oxide with cadmium(II), copper(II), and uranium(IV) demonstrate

that the extent of the displacement to low values of the P-O stretching fre-

quencies is a function of the metal [63].

2. Ultraviolet and Visible Absorption Spectral Studies

The beryllium chelate of H(tta) may exhibit two different solution spectra.

One spectrum resembles the enol form of the β-diketone and has maxima at

350 and 365 mμ [104].

From the absorption spectrum of β-diketonates of uranyl(VI), Socconi

and Giannoni [105] have found that covalent bonds exist between metal and

ligands. The complexes readily tend to hydrate resulting in a coordination

number greater than 6. In a related investigation, the molar absorptivity of

the fluoro-β-diketonates of uranyl(VI) has been found to be maximal (31,200)

with the H(fta) [106]. Spectrophotometric studies of uranyl(VI) complexes

with H(tta) in dilute solution point to a 2:3 complex, while in concentrated

solution a 1:2 complex is indicated [108]. Cotton and Holm [107] have

studied the cobalt(II) complexes of various fluoro-β-diketonates which appear
to be spin-free planar quadricoordinate complexes.

Low-temperature intercombination spectra of crystalline chromium(III)
acetylacetonate and the fluorinated derivatives indicate that the effect of
ligand halogenation on the trigonal splitting is small [111]. In addition,
ligand field spectra of manganese(III) chelates show that these $3d^4$ complexes
are not centrosymmetric at ambient temperatures, but undergo oscillations
from one potential minimum to the other [112].

The electronic spectra of alpha-substituted β-diketonates of copper(II)
show a band at 250 mμ arising from electron transfer from a ligand orbital
to the metal antibonding molecular orbital, while a transition π to π^* located
primarily on the ligand is assigned to a band at 200 mμ [109]. Low-temperature absorption spectra of copper(II) β-diketonates have been studied in the
22,000-40,000 cm^{-1} region for the effects of various ligand substituents.
Shifts in this region are similar for all the bands and are in accord with
molecular orbital theory for $\pi - \pi^*$ transitions [110].

Studies of visible spectra of fluoro-β-diketonates of copper(II) in the
presence of a varying amount of 4-methypyridine indicate a 1:1 chelate base
structure for the chelates [85]. In the presence of tri-n-butyl-, tri-n-octyl-
and triphenylphosphine oxides, these copper(II) chelates yield visible spectrum
data indicating that the ternary complexes formed have electron-donating
ability of the phosphine oxides in the order butyl<phenyl<octyl [63]. In a
related spectrophotometric study using these phosphine oxides as well as
triphenyl phosphate, it is observed that the electron-donor ability increases
in the order triphenyl<tri-n-butyl<triphenoxy<tri-n-octyl-phosphine oxide
[113].

Spectrophotometric monitoring in the visible region of N-base adducts of
copper(II) chelates with quinoline and isoquinoline has been carried out [114].
The effects of axial ligation with pyridine on the ligand field spectrum of
copper(II)hexafluoroacetylacetonate has been studied, and four-band gaussian
analyses presented, wherein features were found to be consistent with
qualitative crystal field predictions [115].

3. Nuclear Magnetic Resonance Studies

Mixtures of tetrakis(trifluoroacetylacetonato) zirconium(IV) and the
tetrakis(acetylacetonato) complex as analyzed by NMR show that in solution,
rapid ligand exchange occurs. Similar exchange is seen for mixtures of
Hf(tfa)$_4$ and Hf(acetylacetone)$_4$ and Zr(tfa)$_4$, Zr(acetylacetone)$_4$ and Zr(dibenzoyl-
methane)$_4$ [116].

On the basis of proton and phosphorus NMR studies, it is reported that
the degree of interaction between H(tta) and tributyl- or tris(butoxyethyl)
phosphate is relatively slight, as is the case with phosphate ester synergists
and zinc(II) in a chelation-extraction process [117]. Apparently the synergism
occurs via exchange of a coordinated water group for the hydrophobic phosphate
ester, thus enhancing extraction into the organic phase.

NMR spectral studies of the mixed chelate systems Cr(III) and Co(III)
(trifluoroacetylacetonate)$_n$(acetylacetonate)$_{3-n}$ where n=1 or 2, in the presence
of excess of the β-diketones indicate that exchange of the CF$_3$ probe between
different chemical environments occurs by an intramolecular mechanism
[118]. In a related study, kinetic parameters for sterochemical rearrange-
ments of trifluoroacetylacetonates of the trivalent metals aluminum, cobalt,
gallium, indium, and rhodium were measured by 19 F NMR [119]. The cis-
trans isomerism was studied and the cis-Rh complex is found to be stable
to isomerization.

Ligand exchange reactions between zirconium and hafnium chelates
derived from acetyl- and trifluoroacetylacetone have been monitored by
NMR in a variety of solvents. The zirconium complexes exchange faster
than the corresponding Hf species derived from acetylacetone, while with
the fluorinated ligand, rates of exchange are essentially the same for both
metals [120, 121].

Phosphorus NMR data for complexes of the type M(tta)$_2 \cdot$ TOPO(M=
divalent metal, TOPO=tri-n-octyl phosphine oxide) have been compiled
[122]. Spectral data was used to determine the equilibrium constant of the

following reaction which occurs in carbon tetrachloride: $Zn(tta)_2 \cdot 2H_2O$ $+2TOPO \rightarrow Zn(tta)_2 \cdot 2TOPO +2H_2O$.

Novel octahedral tin complexes of the type $R_2Sn(L)_2$ (R=methyl, phenyl; L=hexafluoroacetylacetonate) are shown by NMR spectra to consist of the trans configuration with ring proton shifts similar to those seen in most metal β-diketonates [123].

Solvent effects on the NMR spectra of aluminum β-diketonates have been investigated. When applied to aromatic solvents with large diamagnetic anisotropies, it is postulated that in solvents having donor groups, e.g., nitrobenzene, solvation exists along the C_3 axes of the metal chelates [124].

The novel carbon-bonded platinum(II) complexes of the type $K[Pt(tfa)_2X]$ (where X=halogen) have been identified by NMR spectra with both oxygen- and carbon-bonded β-diketone groups present [125]. Rates of optical inversion of $Al(hfa)_3$ by NMR have been measured, and large deviations from the statistical distribution of ligands were found. The mixed ligand complexes were formed in the presence of a second β-diketone. Inversion rates were found to be independent of concentration pointing to an intramolecular racemization process [126].

The methyl and ring proton chemical shifts of the cationic complexes $(C_5H_{12}N)M(tfa)_4$ (M=La, Y) have been ascribed to a charge on the ion rather than a ring current effect [127].

4. Fluorescence and Related Spectral Studies

Rare-earth chelates, dissolved in organic vinylic monomers which were polymerized to solid solutions, give the sharp line fluorescence of f^n orbital transitions which can be attributed to uniform distributions and equivalent surroundings. In the case of the samarium chelate of H(tta), fluorescent lines are found at 645, 598, and 562 mμ at both 77 and 300°K with excitation in a band centered approximately at 360 mμ [128]. The fluorescence spectra of europium and terbium β-diketonates are modified significantly on changing substituents in the ligand [129].

Decay-time determinations for the $^5D_0 \rightarrow {}^7F_2$ transition in the europium(III) thenoyltrifluoroacetonate chelate in methylcyclohexane, ethanol, or toluene at 77°K gave exponential decay curves which were concentration-dependent [130]. Stroboscopic determinations showed an exponential luminescence rise with a rise time of 2 sec.

The temperature dependence of the fluorescence quantum yield and decay time of several rare-earth chelates were studied in an attempt to ascertain the quenching mechanisms in these compounds [131]. Reduction in quenching occurs with lowering the amplitude of molecular vibrations, thus indicating that coupling of the electronic states to the environment through molecular vibrations occurs. The shielded 4f orbitals of rare-earth atoms become quenching-insensitive when they are incorporated into a chelate which indicates some significance in connection with liquid laser research.

An analysis of the spectra of europium(III)thenoyltrifluoroacetonate in organic polymer hosts at different temperatures indicate that the mechanism responsible for the line fluorescence is due to an intramolecular energy transfer from the triplet states of the chelate to the $^5D_0$ and $^5D_1$ resonance levels of the central europium ion [132].

The intramolecular sensitization of the europium(III) hexafluoroacetylacetonate using benzophenone as a sensitizer with excitation at 380 mμ has been examined at room temperature. Mixtures of sensitizer and chelate show some increase in the intensity of the europium red lines [133]. Sublimation of europium(III)hexafluoroacetylacetonate gives a sublimate displaying a fluorescence as strong as the starting material, although high resolution spectra suggest that the sublimate contains more than one species, whereas the original material is one compound [134].

A related study has involved intermolecular triplet-triplet energy transfer between aromatic carbonyls and hydrocarbons as donors, and gadolinium (III) and europium(III) hexafluoroacetylacetonates as acceptors. The transfer is diffusion-controlled and transfer from the donor to the acceptor triplet level is no more efficient than between benzophenone and naphthalene, which indicates that the heavy central ion has little effect on transfer probability [135].

Europium(III) hexafluoroacetylacetonate in EPA solution and in the presence of cis-piperylene shows quenching of chelate fluorescence, while in the presence of naphthalene, europium(III) trifluoroacetylacetonate does not undergo fluorescence quenching [136]. Ligand exchange involving H(hfa) with Sm, Eu, and Tb may be studied by means of fluorescence spectra and indicate that ligand exchange proceeds in the order Tb → Eu → Sm [137]. Similarly time-resolved fluorescence spectra of several europium(III) diketonates show that $^5D_0$ state is populated by nonradiative energy transfer from the higher lying $^5D_1$ state, where relaxation time is measured in the order of microseconds, which suggests that ligand stretching vibrations and bending vibrations of coordinated solvent molecules make important contributions to the relaxation process [138].

The relative luminescence intensities and the mean lifetimes were followed, for the $^5D_0 → ^7F_2$ transitions of europium(III) in its complexes, by thenoyltrifluoroacetylacetone and bases such as phenanthroline, dipyridyl, diphenylguanidine, quinoline, piperidine, collidine, pyridine, and 2-aminopyridine [139]. In all cases, lifetimes were 2.3–3.2 times higher than complexes having coordinated water in place of a base. In the presence of HCl, the luminescence spectra of europium(III) thenoyltrifluoroacetonate shows three patterns corresponding to three complexes in different stages of dissociation, which are dependent on pH [140].

The emission spectra of β-diketonates of europium(III) studied in a variety of solvents having varying ligand concentrations indicate the number of ligands coordinated to europium can vary from one to four [141]. The molecular geometry of the chelate and its effects in emission are also solvent-dependent.

Tetrakis ion-pair chelates of europium derived from H(bta) and an organic base display fluorescence spectra stronger than the neutral aquated

tris chelates, both as solids and in solution, with the nature of the cation
(pyridinium, quinolinium, and isoquinolinium) having a marked effect on the
fluorescent properties [90]. The low quantum efficiencies of a number of
these cation complexes in which europium is octacoordinated have been found,
and are ascribed to strong intramolecular quenching at the organic cation
site [142-144].

5. Mass Spectral Studies

Mass spectral studies have been carried out on a variety of fluoro-β-
diketonates of the first-row transition elements as well as various group III
and IIB ions [145, 146].

The significant factor in the mass spectra of chromium(III) hexafluoroacetylacetonate derivative is the tendency toward formation of even-electron
ions, accompanied by metal valence change where necessary. The initial
radical cation(molecular ion) is formed as a result of electron removal from
the ligand pi system. Fragmentation proceeds with loss of neutral ligand
and retention of metal charge on the remaining even-electron ion, the latter
suffering loss of a second neutral ligand with a one-electron reduction of the
central metal ion. Alternative processes in which metal reduction occurs
are loss of an alpha or gamma substituent as an odd electron neutral fragment. Fluorine substitution appears to favor loss of an alpha substituent
over the complete departure of a complete neutral fragment. In the mass
spectrum of the hexafluoroacetylacetonate, double-charged ions are observed.

D. Analytical Applications

1. Solvent Extraction Studies of Various Ions Employing Fluoro-β-Diketones

The experimental data for extraction of the various classes of metal ions employing fluoro-β-diketones are given in Table 8. Emphasis is given to those reports which describe the optimal conditions for extraction of a given ion. Theoretical treatments of solvent extraction procedures and related reports are outside of the intent of this review and are not included.

Table 8 consists of the ion extracted, the agent employed, the organic phase, the aqueous phase, and the corresponding reference.

2. Gas Chromatography of Fluoro-β-Diketonates

In the last few years the use of gas chromatography has revolutionized research in organic chemistry and biochemistry. The properties that make gas-liquid chromatography so attractive are great sensitivity, the speed and ease of separating mixtures of closely related compounds, and the easy rapid measurement of thermodynamic data. The most significant areas in which gas chromatography is useful in the study of metal complexes are investigations of sterochemistry, ligand exchange, isomerization, metal-ligand stoichiometry, interactions of complexes with weak donors, separation of geometrical and optical isomers, and ultratrace metal analysis [45]. In order to achieve gas-chromatographic separations, the metal complex must be volatile and thermally and solvolytically stable in the chromatographic column. The β-diketonates

$$\begin{array}{c} M \\ \diagup \quad \diagdown \\ O \qquad O \\ | \qquad | \\ R_1-C-CH-C-R_2 \end{array}$$

exhibit the properties where

$$R_1=CF_3, R_2=CH_3 \quad \text{(tfa)}$$
$$R_1=CF_3, R_2=CF_3 \quad \text{(tfa)}$$
$$R_1=CF_3CF_2CF_2, \ R_2=C(CH_3)_3 \quad \text{(fod)}$$
$$R_1=CF_3CF_2CF_2, \ R_2=CF_3 \quad \text{(dfhd)}$$
$$R_1=C(CH_3)_3, \ R_2=CF_3 \quad \text{(dapm)}$$

TABLE 8

Solvent Extraction Studies of Various Ions Employing Fluoro-β-Diketones

Ion(s) extracted	Organic phase	Aqueous phase	Ref.
I. Thenoyltrifluoroacetone			
A. Non-rare-earth ions			
Ac, Be, Bi, Hf, In, Fe, Np, Pb, Pu, Ra, Sc, Tl, Th, U, V, Y	Benzene	Buffered solutions, variable pH	[147]
Ac	Benzene	pH 3.5-5.0	[148]
Ac	Benzene	Variable pH	[149]
Al, Ba, Ca, Ga, In, Hf, Mg, Sc, Sr, Th, U, Y, Zr	Chloroform	Buffered solutions, variable pH	[150]
Al	Hexone	pH 5.5-6.0(acetate), p pH 2.5-4.5(cupferron)	[151]
Am	Chloroform + dibutyl phosphate	Buffered	[152]
Am	Hexone or chloroform	0.1 M H_3O^+ or $NaClO_4$	[153]

TABLE 8 (Continued)

Solvent Extraction Studies of Various Ions Employing Fluoro–β–Diketones

Ion(s) extracted	Organic phase	Aqueous phase	Ref.
Am, In, Sc	Carbon tetrachloride + tributyl phosphate, dibutylsulfoxide or hexone	1.0 M H_3O^+, $NaClO_4$	[154]
Am	Benzene	pH 3.23	[155]
Am	Carbon tetrachloride + tributyl phosphate	4 M $NaClO_4$, 0.1 M H_3O^+ and variable Cl	[156]
Am	Carbon tetrachloride + hexone	1 M $NaClO_4$, variable $SO_4^=$	[157]
Am	Benzene	Solutions of oxalic acid, tartaric acid, or nitrilotriacetic acid	[158]
Am	Benzene	0.1 M NH_4Cl	[159]
Am, Cm	Benzene	Variable pH	[160]

Ba	Carbon tetrachloride + tributyl phosphate	1.0 M NaClO$_4$	[161]
Ba, Ca, Mg, Sr	Hexone	NH$_4$OAc–AcOH–NH$_4$OH	[162]
Be	Hexone	Sulfanilate buffer + SO$_4^=$, ClO$_4^-$, or oxalate	[163]
Bk	Xylene	0.5–3.5 N HNO$_3$, 0.5–1.0 N H$_2$SO$_4$, or 0.1 N HCl	[164]
Ca	Carbon tetrachloride + tri-n-octyl phosphine oxide, tributyl phosphate, or hexone		[165]
Co, Np	Various alcohols, ketones or esters	Variable pH	[166]
Co, Ni, Pb, Zn	Toluene	Weak acids	[167]
Co	Benzene	Variable pH, NaClO$_4$	[168]
Co, Fe, In, Sc, U, Zn	Chloroform, carbon tetrachloride, or benzene	Variable pH	[169]
Co	Acetone–benzene	Ammonium acetate buffer	[170]

TABLE 8 (Continued)

Solvent Extraction Studies of Various Ions Employing Fluoro-β-Diketones

Ion(s) extracted	Organic phase	Aqueous phase	Ref.
Co	Cyclohexane + heterocyclic base	Variable pH	[171]
Cr	Benzene	Variable pH	[172]
Cs	Nitrobenzene	pH 8.7–9.0 LiOH or 1 N Na$_2$CO$_3$	[173]
Cs, Rb	Nitrobenzene	0.06 N NaOH	[174]
Cu	Benzene + variable heterocyclic bases	Acetate buffered	[114]
Cu, Zn	Variable solvents with -isopropyltropolone + phosphorus esters or a heterocyclic base	Buffered	[175]
Cu	Benzene + quinoline, isoquinoline, or trioctyl phosphine oxide	Buffered	[176]

Cu, U	Benzene + two β-diketones	Buffered	[177]
Cu	Variable organic solvents + tributyl phosphate	Buffered	[178]
Cu	Benzene	pH 2.4–6.0	[179]
Cu, Zn	Carbon tetrachloride, chloroform or hexone + tributyl phosphate, hexone or isopropyltropolone	0.1 M H_3O^+, $NaClO_4$	[180]
Cu, Zn	Carbon tetrachloride, chloroform or hexone + isopropyltropolone	0.1 M H_3O^+, $NaClO_4$	[181]
Fe	Benzene	pH 2.0	[182]
Fe	Xylene	2.0 M HNO_3 or 9.0 M NH_4NO_3	[183]
Fe	Variable	10.0 M HNO_3	[184]
Fe	Benzene	$HClO_4$ + SCN^-	[185]
Fe, Mn	Xylene	0.5 M H_2SO_4 + 0.3 M $NaBrO_3$	[186]

TABLE 8 (Continued)

Solvent Extraction Studies of Various Ions Employing Fluoro-β-Diketones

Ion(s) extracted	Organic phase	Aqueous phase	Ref.
Fe, U	Benzene or nitrobenzene + tri-n-octylamine	0.5 M HCl (Fe), 2.0 M HNO$_3$ (U)	[187]
Hf, Zr	Benzene	2.0 N HClO$_4$	[188]
Hf	Benzene	Cl$^-$, F$^-$, NO$_3^-$, or SO$_4^=$	[189]
Hf	Benzene	3.0 M HClO$_4$ + F$^-$	[190]
Hf	Benzene	2.0 M H$_3$O$^+$	[191]
Hf	Benzene	=1.0	[192]
Hf, Zr	Benzene or o-dichlorobenzene	4.0 M HClO$_4$	[193]
Hf, Zr	Benzene	HClO$_4$ or HCl	[194]
In	Chloroform + -isopropyl-tropolone	ClO$_4^-$	[195]
Mg, Ca, Sr, Ba	Variable + organophosphorus synergists	Variable pH	[196]

Mn	Benzene + pyridine	Acidic, then neutralized to bromcresol green	[197]
Mn	Benzene-acetone	pH 6.7-8.0	[198]
Mn	Xylene	0.5 M H_2SO_4 + BrO_3^-	[199]
Mo, Np	Extraction chromatography tri-n-octylamine/xylene		[200]
Mo, W	Butanol-acetophenone	3.0-9.0 N HCl (Mo) 9.0-10.0 N HCl (W) + $SnCl_2$ + SCN^-	[201]
Nb	Xylene + n-butanol	Conc HCl or HCl/H_2SO_4	[202]
Nb, Zr	Xylene	Variable conc HCl	[203]
Nb, Zr	Xylene		[204]
Ni	Acetone-benzene	Buffered, pH 5.5-8.0	[205]
Np	Benzene or xylene		[206]
Np	Benzene, toluene, or hexa-fluoroxylene	Variable pH	[207]
Np	Xylene/tri-iso-octylamine	5.0 M HNO_3	[208]
Np	Butanol and other alcohols	pH 7.0-10.0	[209]

TABLE 8 (Continued)

Solvent Extraction Studies of Various Ions Employing Fluoro–β–Diketones

Ion(s) extracted	Organic phase	Aqueous phase	Ref.
Np	Benzene	Cl^- or NO_3^-	[210]
Np	Benzene	0.05–2.0 M HCl	[211]
Np, U	Isobutanol, cyclohexanone butanol, MeEtCO–amyl or ethyl acetate, hexyl or isoamyl alcohol	pH 5.3–9.3	[212]
Pa	Benzene	6.0 N HCl	[213]
Pa	Benzene, hexafluoroxylene, or toluene	0.05–16.0 N HCl	[214]
Pa	Benzene	6.0 N HCl	[215]
Pa	Benzene	1.8–9.0 N HCl	[216]
Pa	Benzene	HCl, HNO_3, or H_2SO_4	[217]

Element	Solvent	Medium	Ref.
Pa	Benzene	HCl	[218]
Pa	Benzene	$HClO_4$–$LiClO_4$	[219]
Pa	Benzene	$HClO_4$–$LiClO_4$	[220]
Pa	Benzene	6.0 M HCl	[221]
Pa	Benzene	$HClO_4$–$LiClO_4$	[222]
Pa	Benzene	Dilute $HClO_4$	[223]
Pa	Benzene	Variable	[224]
Pd, Pt	Acetophenone–butanol	5–9 N HCl + $SnCl_2$ (Pt)	[225]
Pu	Benzene	0.5–1.0 N H_3O^+	[226]
Pu	Benzene	HNO_3	[227]
Pu	Benzene	ClO_4^-, NO_3^-, or Cl^-	[228]
Pu, U	Benzene	=4.1	[229]
Pu	Benzene	8.0 M HNO_3	[230]
Pu, V	Cyclohexane + tributyl phosphate or tributyl phosphine oxide	HNO_3	[231]
Pu	Benzene	Variable	[232]
Pu, U	Variable	Variable pH	[233]

TABLE 8 (Continued)

Solvent Extraction Studies of Various Ions Employing Fluoro-β-Diketones

Ion(s) extracted	Organic phase	Aqueous phase	Ref.
Rh	Acetone-xylene	pH 6.0 + ClO_4^-	[234]
Sc	Benzene + polar solvent		[235]
Sc	Variable	$NaClO_4$ (μ = 0.1)	[236]
Sc	Benzene	pH 1.5	[237]
Sc	Chloroform	0.1 M H_3O^+, $NaClO_4$	[238]
Sc, Y	Benzene	Low pH	[239]
Sc, Ag, Cd, Cu, Fe, Mn, Ni, Pb, Pd, Co, Hg, Th	Hexone, butyl acetate, or isoamyl acetate	Buffered	[240]
Sn	Hexone	$H_4SO_4-Cl^-$	[242]
Sn	Variable	Variable pH	[243]
Sr	Hexone	pH 9.5	[244]
Sr	Benzene	1.0 N HCl or HNO_3	[245]

	Solvent	Conditions	Ref.
Sr, Y	Benzene	Variable pH	[246]
Sr, Y	Benzene	pH 6-9	[247]
Tc	Benzene	Variable pH	[248]
Th	Benzene or cyclohexane		[249]
Th	Benzene	Variable pH	[250]
Th	Benzene	HNO_3 (pH 2.5)	[251]
Ti	Benzene	0.1 M ClO_4^-	[252]
Ti	Benzene	pH 0.5–2.0	[253]
Ti	Benzene-isoamyl alcohol	10.0 M HCl	[254]
Tl	Chloroform	Variable pH, 0.1 M $NaClO_4$	[255]
Tl	Variable	Buffered	[256]
Tl, Zn	Variable	pH 0–7	[241]
U	Benzene + tri-n-octylamine	$HClO_4$ (pH 0.8)	[258]
U	Benzene	Variable	[259]
U	Benzene	HNO_3, Variable pH	[260]
U	Benzene	pH 3.0	[261]
U	Benzene	Variable pH, $HClO_4$	[262]
U	Carbon tetrachloride	0.001 N HCl, H_2SO_4, or HNO_3	[263]

TABLE 8 (Continued)

Solvent Extraction Studies of Various Ions Employing Fluoro-β-Diketones

Ion(s) extracted	Organic phase	Aqueous phase	Ref.
U	Variable	0.01 M acetate, pH 1–6	[264]
U	Benzene + tri–n–butyl phosphate	2.0 M LiClO$_4$	[265]
U	Benzene + oxine	Acetate buffered, pH 3.5	[266]
V	Variable		[267]
Y	Benzene, chloroform	pH 4.0	[268]
Zn	Variable + tri–n–octyl phosphine oxide	pH 4.6	[269]
Zr	Benzene	2.0 M HNO$_3$ + urea	[270]
Zr	Chloroform	0.7–3.0 M HNO$_3$	[271]
Zr	Variable		[272]

Zr	Benzene	2.0 M HClO$_4$	[273]
Zr	Benzene, chloroform	Variable	[274]
Zr	Benzene	Variable	[275]
B. Rare-earth ions			
Ce	Xylene	1.0 N H$_2$SO$_4$	[276]
Ce	Benzene	Variable pH	[250]
Ce	Xylene	1.0 N H$_2$SO$_4$	[277]
Ce	Benzene + tri-n-butyl phosphate	Variable pH	[278]
Ce	Benzene	pH 5.4	[279]
Ce, Dy, Eu, Gd, Ho, La, Nd, Pr, Pm, Sm, Tm	Benzene	Variable pH	[147]
Ce, Nd, Pm, Sm	Benzene	pH 4.0	[280]
Ce	Xylene	0.5 M H$_2$SO$_4$ -0.3 M BrO$_3^-$	[186]
Ce	Variable alcohols, esters, or ketones	Variable pH	[166]
Er, Dy, Ho, Tm, Yb	Benzene	pH 6.5	[281]
Eu	Toluene	Oxalate, Cl$^-$ or ClO$_4^-$	[282]

TABLE 8 (Continued)

Solvent Extraction Studies of Various Ions Employing Fluoro-β-Diketones

Ion(s) extracted	Organic phase	Aqueous phase	Ref.
Eu	Toluene	Acetate, glycolate, or sulfate	[283]
Eu	Chloroform + tributyl phosphate	0.1 M NaClO$_4$	[284]
Eu	Benzene + oxine	Acetate buffered, pH 3.5	[266]
Eu	Chloroform + -isopropyl-tropolone	ClO$_4^-$	[195]
Eu	Chloroform + dibutyl phosphate	Buffered	[152]
Eu	Hexone or chloroform	0.1 M H$_3$O$^+$, NaClO$_4$	[153]
Eu, Lu	Benzene	Low pH	[239]
Eu, Tb	Benzene + 2-hexone	pH 3.7, acetate + NH$_4$OAc	[285]
Eu, Sm	Benzene + collidine or diphenylguanidine	pH 6.5-7.5	[286]

Eu, Cd, La, Y, Yb	Benzene + tributyl phosphate	0–8 M LiNO$_3$, pH 1–7	[287]
Eu, La, Lu	Carbon tetrachloride + tributyl phosphate, dibutylsulfoxide, or hexone	1.0 M H$_3$O$^+$, NaClO$_4$	[154]
Eu, La, Lu	Carbon tetrachloride + tributyl phosphate, or hexone	1.0 M NaClO$_4$, variable SO$_4$	[157]
Ho, Lu, Pr	Benzene		[288]
La	Benzene	pH 3.23	[155]
Nd	Chloroform	Variable pH	[150]
Pm	Benzene	0.1 M NH$_4$Cl	[159]
Pm	Benzene	Oxalic acid, tartaric acid, or nitrilotriacetic acid	[158]
Pr	Benzene	pH 3.0–4.2	[289]
II. Trifluoroacetylacetone			
Ac	Benzene	Variable pH	[149]

TABLE 8 (Continued)

Solvent Extraction Studies of Various Ions Employing Fluoro-β-Diketones

Ion(s) extracted	Organic phase	Aqueous phase	Ref.
Al, Cr, Mg, Mn	Chloroform	0.25–2.5 N NH$_4$CL, 0.1 N NaOH or 1.0 M acetate	[290]
Be	Benzene, chloroform, butanol, or toluene	pH 1–1.5 + EDTA	[291]
Cd, Co, Cu, Ni, Pd, Zn	Chloroform + isobutylamine	pH 9.0	[292]
Cu	Benzene + variable heterocyclic bases	Acetate buffered	[114]
Cu	Benzene + quinoline, isoquinoline, or trioctyl phosphine oxide	Buffered	[176]
Eu	Chloroform + tributyl phosphate	0.1 N NaClO$_4$	[284]

Element	Solvent	Conditions	Ref.
Hf, Zr, Zn	Benzene	0.2 N HCl	[293]
Np	Benzene	0.05–2.0 M HCl	[211]
Np, Pu, U, Zr	Benzene, toluene, or hexafluoroxylene	Variable pH	[207]
Pa	Benzene, hexafluoroxylene, or toluene	0.05–16.0 N HCl	[214]
Sc	Benzene + polar solvent		[235]
Sc	Variable	$NaClO_4\,(\mu=0.1)$	[236]
Zr	Variable		[272]
III. Hexafluoroacetylacetone			
Be	Benzene, butanol, chloroform, or toluene	pH 1–1.5 + EDTA	[291]
Cu	Benzene + variable heterocyclic bases	Acetate buffered	[114]
Cu	Benzene + quinoline, isoquinoline, or trioctyl phosphine oxide	Buffered	[176]

TABLE 8 (Continued)

Solvent Extraction Studies of Various Ions Employing Fluoro-β-Diketones

Ion(s) extracted	Organic phase	Aqueous phase	Ref.
IV. Benzoyltrifluoroacetone			
Eu	Chloroform + tributyl phosphate	0.1 N NaClO$_4$	[284]
Hf, Zr	Benzene or o–dichlorobenzene	4.0 N HClO$_4$	[193]
Np	Benzene	0.05–2.0 N HCl	[211]
Np, Pu, U, Zr	Benzene, toluene, or hexafluoroxylene	Variable pH	[207]
Pa	Benzene, hexafluoroxylene, or toluene	0.05–16.0 N HCl	[214]
U	Benzene + tributyl phosphate	2.0 N LiClO$_4$	[265]
U	Butyl acetate	Buffered	[257]
Zr	Variable		[272]

V.	(Selenofuroyl) trifluoroacetone			
	Hf, Zr	Benzene	1.0 N HClO$_4$	[294]
	Hf	Benzene	μ=1.0	[192]
	Nd	Chloroform		[295]
VI.	R-CO-CH$_2$-CO-R' (R=cycloalkyl, halogenated aryl; R'=CF$_3$)			
	Np, Ga	Variable aromatics	Variable pH	[296]

It is now well established that the volatility of complexes and the ease with which they are eluted is greatly increased by substitution of fluorine for hydrogen in the "ligand" shell [312, 313].

Figure 1 gives a summary of some of the gas chromatographic studies of various fluoro-β-diketonates. Moshier and Sievers [45] as well as Juvet and Zado [312] have reviewed gas-chromatographic studies in great detail. In addition, radioactive tracers have been employed to determine if samples decompose or adsorb on the injection port or the chromatographic column [297, 298, 300, 316].

Using Fig. 1, examples of volatile complexes of almost every class of metals can be shown. For many years the complexes of the lanthanides that are volatile and stable have been of particular attention. Examination of the chromatographic data for complexes of the lanthanides reveal that complexes of lanthanides with larger ionic radii are eluted much later than those with smaller radii [308]. This same volatility trend is shown to be exhibited by other classes of chelates [310, 46, 320-323].

Gas chromatography has been found to be most useful in the studies of stereochemistry [56, 310, 311, 61]. The technique is especially effective with mixtures of isomers possessing only subtle differencies in structure. Palmer et al. [118], have demonstrated that mixed-ligand complexes can easily be separated and analyzed by gas chromatography. Data of this type should prove to be extremely useful in studies of the kinetics and equilibria of isomerization, ligand substitution, and ligand-exchange reactions.

The separation of optical isomers presents a unique problem since dextro and levo isomers have identical dipole moments, vapor pressures, etc. Unless an asymmetric system exists within the column, no separation is possible. However, this asymmetric environment is possible when an optically active liquid is employed as the stationary phase. Sievers et. al. [305] were able to achieve partial separation of the optical isomers of Cr(hfa)$_3$ by gas-solid chromatography, employing an optically active solid absorbent in the column. Also it has been proposed that it might be possible to use gas-chromatographic retention measurements as a means to determine

GAS CHROMATOGRAPHY OF FLUORO-β DIKETONATES

Group 2	Group 3	Ti	V	Cr	Mn	Fe	Co	Ni	Cu	Zn	Group 13
Be tfa (297, 299, 56, 311, 313-4); hfa (299); fod (304)											**Al** tfa (299, 56, 305, 307, 311, 313-5); hfa (299, 301); fod (304)
Mg tfa (311)											
Ca tfa (45, 313,-4); fod (308); tapm (310)	**Sc** tfa (45, 313,-4); fod (308); tapm (310)	**Ti (IV)**	**V (IV)** tfa (45)	**Cr(III)** tfa (301-3, 56, 309, 313); hfa (298-301, 56, 305, 311, 313-4); fod (304)	**Mn(III)** tfa (311, 313, 314)	**Fe(III)** tfa (299, 56, 305, 311, 313, 315); hfa (298); fod (304)	**Co(III)** tfa (74, 311); hfa (74)-	**Ni(II)** fod (304)	**Cu(II)** tfa (299, 302, 56, 305, 311, 313-5); hfa (313); fod (304)	**Zn(II)** tfa (311, 313-314)	**Ga** tfa (307, 311, 313-4)
Sr fod (304, 308); tapm (306)	**Y** fod (304, 308); tapm (306)	**Zr** tfa (56, 313-4); hfa (313)	**Nb** hfa (313)	**Mo**	**Tc**	**Ru(III)** tfa (74, 309); hfa (74)	**Rh(III)** tfa (56, 303, 311, 313); hfa (56, 298)	**Pd** fod (304)	**Ag**	**Cd**	**In** tfa (56, 304, 307, 311, 313-314)
Ba	**57-71**	**Hf** tfa (56, 313-4); hfa (313)	**Ta** hfa (313)	**W**	**Re**	**Os**	**Ir**	**Pt**	**Au**	**Hg**	**Tl** tfa (313-4)
Ra	**89-103** Th-tfa (311, 314)										

Ce	Pr	Nd	Pm	Sm	Eu	Gd	Tb	Dy	Ho	Er	Tm	Yb	Lu
		tfa (313); tapm (310)		fod (308); tapm (306, 310)	fod (308); tapm (306)	fod (308); tapm (306)	fod (308); tapm (306)	fod (308); tapm (306, 310)	fod (308); tapm (306)	fod (308); tapm (306, 310)	fod (308); tapm (306)	fod (308); tapm (306)	fod (308); tapm (306, 310)

Fig.1. Gas-chromatographic studies of selected metal fluoro-β-diketonates. (The numbers in parentheses refer to references at the end of the chapter.)

absolute configurations of closely related optically active compounds [317–319].

3. Measurement of Metals by Gas Chromatography

The measurement of metals by gas chromatography was originally considered a novelty and an academic exercise, but it is now evident that the application of this technique can be a valuable tool complementing existing methods of metal analysis. A volatile derivative, of course, must be prepared from the metal prior to analysis, and either metal chelates or metal halides have been employed. The fluoro-β-diketone chelates have been the most extensively studied class of compounds due to their high volatility and excellent thermal and solvolytic stability.

The earlier analytical methods used samples which were dissolved in aqueous media. Following adjustment of the pH, the metal chelates were prepared from the aqueous solution and extracted into an organic solvent. Aliquots of the organic extract then were injected onto the column of the gas chromatograph.

Morie and Sweet [307] in 1965, described the analysis of mixtures of aluminum, gallium, and indium. These metals were extracted from an aqueous medium by shaking for 4 hr at room temperature with a solution of H(tfa) in benzene. Gas-chromatographic measurements were carried out using a stationary phase of silicone DC-550 oil and a thermal conductivity detector. Morie and Sweet [324] also measured aluminum and iron in a nickel-copper alloy using initial dissolution in hot concentrated perchloric acid. The extraction and gas-chromatographic procedures were similar to those described in their earlier paper.

Moshier and Schwarberg [315] similarly employed gas chromatography in the analysis of National Bureau of Standards alloys and quantitatively determined aluminum, iron, and copper. The procedure involved solution of the alloy, conversion of the metal ions to trifluoroacetylacetonates by extraction at room temperature with H(tfa) in chloroform and, finally, complete gas-chromatographic separation of the metal chelates using a polyethylene wax as a stationary phase and a thermal conductivity detector. A similar method [297,325] was employed for the determination of beryllium in aqueous solution. In this case the sensitivity was increased by the use of an electron-capture detector in the gas-chromatographic measurements. The extraction of the metal from aqueous solution was accomplished by shaking for 1 hr at room temperature with a 0.005 M solution of H(tfa) in benzene. Unreacted ligand was then removed by washing with 0.01 M sodium hydroxide. This alkaline wash did not affect the Be(tfa)$_2$ and was of prime importance in preparing the reaction mixture for gas chromatography with electron-capture detection. Without the wash there was sufficient excess ligand to completely saturate and poison the detector. The lower limit of detectability using this method was approximately 4×10^{-13} g of beryllium. A similar procedure [316] was employed for the measurement of trace concentrations of aluminum in uranium. Ross and Sievers [326] measured traces of chromium in ferrous alloys by reacting the alloy with H(tfa) in an rf field. Detection of the Cr(tfa)$_3$ was accomplished using SE-52 as the stationary phase with electron-capture detection. These same workers [304] introduced H(fod) as a chelating agent and found that they could react the neat ligand directly with several metals in a sealed glass capillary tube. The efficacy of the technique was tested by applications to the measurement of iron in Mesabi iron ore.

An important new application of gas chromatography of metal fluoro-β-diketonates has been directed toward the measurement of trace metals in biological materials. Of the 37 elements present in the human body, 26 are metallic in nature and many are present in only trace quantities. The importance of several of these trace metals in mediating biochemical processes has prompted investigations of their physiological values and alterations of these values in pathological states. Of particular significance is the relationship between trace metals and cancer since it is known that many metals such as iron, aluminum, chromium, cobalt, lead, nickel, selenium, and beryllium have carcinogenic properties. In addition, a considerable number of studies have been carried out to determine the function of chromium in biological systems, especially the relationship of low chromium states to abnormal carbohydrate metabolism. The occurrence and function of chromium in biological systems has been reviewed recently [327].

The first report of the application of gas chromatography to trace-metal determination was made in 1968 from the authors' laboratory [328]. In this report, toxic levels of chromium were measured in serum samples. Extensions of the method subsequently were reported [329, 330] and included the measurement of chromium in urine [330]. The biological sample was first digested by heating with a mixture of concentrated mineral acids to decompose the organic matter of the sample. After buffering to pH 6.0, chromium was isolated from the digested samples by chelation extraction with 0.3 M H(tfa) in benzene, a process which involved shaking at 70°C for 45 min. After removal of excess chelating agent by alkaline washing, the benzene solution of $Cr(tfa)_3$ was analyzed by gas chromatography employing a QF-1 stationary phase and electron-capture detection. The gas-chromatographic detection system was capable of detecting 0.03 pg of chromium as $Cr(tfa)_3$. Subsequently a method has been developed [331] for the quantitative determinations of chromium in blood bypassing the mineral-acid digestion step. Chromium was chelated and extracted from blood as $Cr(tfa)_3$ by direct reaction with a hexane solution of H(tfa) in a sealed tube at 150-175°C. After removal of the unreacted ligand by alkaline washing, the $Cr(tfa)_3$ was determined by gas-chromatography and electron-capture detection. Chromium was determined at the 5 μg/100 ml level in blood. The present authors [332] modified Hansen's procedure by extracting chromium from a protein precipitate of serum, and by this approach were able to measure precisely the chromium levels in normal and pathological serum samples.

Similar techniques have been applied to the measurement of beryllium in biological samples. The importance of these measurements arise from the high toxicity of this metal at low concentrations and, thus, the extreme sensitivity of gas chromatography with electron-capture detection has been of value. The need for beryllium measurements has resulted from the widespread application of beryllium in the fluorescent tube industry and in metallurgical and rocket propellant technology. In one method [333] for the measurement of beryllium, the biological sample (blood, homogenized plant leaf, or liver), was extracted with a benzene solution of H(tfa) in a sealed tube at 120°C. After removal of excess ligand by NH_4OH, the $Be(tfa)_2$ was measured by gas chromatography. A typical analysis took less than 1 hr and as little as 2.95 x 10^{-7} g of beryllium per 50 μl sample was routinely determined. A more complex procedure [334] was described for measuring beryllium in urine, blood, tissue, and air-borne dust. The method involved an initial wet diges-tion for biological samples or ashing in a muffle furnace for dust. The digest was adjusted to pH 5-6 and $Be(tfa)_2$ was extracted by shaking the solution at room temperature for 2 hrs with H(tfa) in benzene. As before, excess H(tfa) was removed by alkaline washing and 1-5 μl of the organic extract was injected onto the gas-chromatographic column. Levels as low as 0.0001 μg of beryllium could be detected.

These preliminary reports indicate the immense potential of gas chromato-graphy of metal fluoro-β-ketonates as an analytical tool for measuring trace-metal concentrations. Contributions of this technique to medicine and environmental control could prove to be the most valuable aspect of the chemistry of the fluoro-β-diketonates.

REFERENCES

[1] S. A. Chaston, S. E. Livingston, T. N. Lockyer, V. A. Pickles, and J. S. Shannon, Australian J. Chem., 18, 673 (1965).

[2] R. K. Y. Ho, S. E. Livingstone, and T. N. Lockyer, Australian J. Chem., 19, 1179 (1966).

[3] R. K. Y. Ho, S. E. Livingstone, and T. N. Lockyer, Australian J. Chem., 21, 103 (1968).

[4]   R. K. Y. Ho and S. E. Livingstone, Australian J. Chem., 21, 1781
      (1968).

[5]   E. W. Berg and K. P. Reed, Anal. Chim. Acta, 36, 372 (1966).

[6]   A. Gero, J. Org. Chem., 19, 469 (1954).

[7]   J. D. Park, H. A. Brown, and J. R. Lacher, J. Am. Chem. Soc.,
      75, 4753 (1953); and references cited therein.

[8]   J. L. Burdett and M. T. Rogers, J. Am. Chem. Soc., 86, 2105
      (1964).

[9]   C. R. Hauser, F. W. Swamer, and J. T. Adams, in Organic Reactions,
      (R. Adams et al., eds.) Vol. VIII, Wiley, New York, p. 59.

[10] A. L. Henne, M. S. Newman, L. L. Quill, and R. A. Staniforth,
      J. Am. Chem. Soc., 69, 1819 (1947).

[11] H. I. Schlesinger and H. C. Brown, U.S. Pat. 662,600 (1951).

[12] J. C. Reid and M. Calvin, J. Am. Chem. Soc., 72, 2948 (1950).

[13] R. Levine and J. K. Sneed, J. Am. Chem. Soc., 73, 4478 (1951).

[14] L. B. Barkley and R. Levine, J. Am. Chem. Soc., 73, 4625 (1951).

[15] L. B. Barkley and R. Levine, J. Am. Chem. Soc., 73, 2059 (1953).

[16] R. A. Moore and R. Levine, J. Org. Chem., 29, 1883 (1964).

[17] E. D. Bergmann, S. Cohen, and I. Shohak, J. Chem. Soc., 1961,
      3278.

[18] J. P. Fackler, Jr., and F. A. Cotton, J. Chem. Soc., 1960, 1435.

[19] C. E. Inman, R. E. Oesterling, and E. A. Tyezkowski, J. Am.
      Chem Soc., 80, 6533 (1958).

[20] J. Stary, The Solvent Extraction of Metal Chelates, Pergamon,
      Oxford, 1964; and references cited therein.

[21] M. Calvin and K. W. Wilson, J. Am. Chem. Soc., 67, 2003 (1945).

[22] R. L. Belford, A. E. Martell, and M. Calvin, J. Inorg. Nucl. Chem.,
      2, 11 (1956).

[23] B. G. Schultz and E. M Larsen, J. Am. Chem. Soc., 71, 3250 (1949).

[24] K. Sato, Y. Kodama and K. Arakawa, Nippon Kagaku Zasshi, 87, 821
      (1966).

[25] K. Sato and K. Arakawa, Nippon Kagaku Zasshi, 88, 470 (1967).

[26] K. Sato, Y. Kodama, and K. Arakawa, Nippon Kagaku Zasshi, 87, 1092 (1966).

[27] E. H. Cook and R. W. Taft, Jr., J. Am. Chem. Soc., 74, 6103 (1952).

[28] D. M. Brower, Chem. Commun., 1967, 515.

[29] G. A. Olah and C. U. Pittman, Jr., J. Am. Chem. Soc., 88, 3310 (1966).

[30] P. J. Elving and C. M. Callahan, J. Am. Chem. Soc., 77, 2077 (1955).

[31] P. J. Elving and P. G. Godzka, U.S. Atomic Energy Comm., AE-U-3960, (1958).

[32] P. J. Elving and P. G. Grodzka, Anal. Chem., 33, 2 (1961).

[33] E. Sawicki and V. T. Oliverio, J. Org. Chem., 21, 183 (1956).

[34] W. B. Whalley, J. Chem. Soc., 1951, 3235.

[35] M. Lipp, F. Dallacke, and S. Munner, Ann., 618, 110 (1958).

[36] H. A. Wagner, U.S. Pat. 3,200,128 (1965).

[37] S. A. Fuqua and R. M. Silverstein, Chem. Ind. (London), 1963, 1591.

[38] R. G. Haber and E. Schoenberger, Neth. Appl., 6,504,329 (1965).

[39] K. J. Rorig, U.S. Pat. 2,748,119 (1956).

[40] A. Prakash and I. R. Gambhir, J. Indian Chem. Soc., 43, 529 (1966).

[41] E. M. Larsen and G. A. Terry, J. Am. Chem. Soc., 73, 500 (1951).

[42] T. R. Norton, U.S. Pat. 3,362,935 (1968).

[43] R. N. Haszeldine, W. K. R. Musgrave, R. Smith and L. M. Turton, J. Chem. Soc., 1951, 609.

[44] W. G. Scribner, Annual Report, Contract No. AF 33(615)-1093, September, 1967.

[45] R. W. Moshier and R. E. Sievers, The Gas Chromatography of Metal Chelates, Pergamon, New York, 1965.

[46] C. S. Springer, Jr., D. W. Meek, and R. E. Sievers, Inorg. Chem., 6, 1105 (1967).

[47] B. P. Pullen, M.S. dissertation, University of Tennessee, 1967.

[48] G. K. Schweitzer, B. P. Pullen, and Y. H. Fang, Anal. Chim. Acta, 43, 332 (1968).

[49] H. I. Schlesinger, H. C. Brown, J. J. Katz, S. Archer, and R. A. Lad, J. Am. Chem. Soc., 75, 2446 (1953).

[50] R. G. Charles and E. P. Riedel, J. Inorg. Nucl. Chem., 29, 715 (1967).

[51] P. A. Whittaker and E. R. Redfearn, Biochem. J., 88, 15P (1963).

[52] S. Takemori and T. E. King, Science, 144, 852 (1964).

[53] E. R. Redfearn, P. A. Whittaker, and J. Burgos, Proc. Symp. Oxidases Related Redox Systems, Amherst, Mass., 1964.

[54] A. Furst, W. C. Cutting, and R. H. Dreisbach, Stanford Med. Bull., 12, 190 (1954).

[55] R. A. Staniforth, doctoral dissertation, Ohio State University, 1943.

[56] R. E. Sievers, B. W. Ponder, M. L. Morris, and R. W. Moshier, Inorg. Chem., 2, 693 (1963).

[57] M. L. Morris, R. W. Moshier, and R. E. Sievers, Inorg. Syn., 9, 50 (1967).

[58] T. G. Dunne and F. A. Cotton, Inorg. Chem., 2, 263 (1963).

[59] E. R. Melby, N. J. Rose, E. Abrumson and J. C. Caris, J. Am. Chem. Soc., 86, 5117 (1964).

[60] L. J. Boucher and J. C. Bailar, Inorg. Chem., 3, 589 (1964).

[61] R. C. Fay and T. S. Piper, J. Am. Chem. Soc., 85, 500 (1963).

[62] E. W. Berg and J. T. Truemper, J. Phys. Chem., 64, 487 (1960).

[63] G. N. Rao and N. C. Li, Can. J. Chem., 44, 2775 (1966).

[64] D. A. Buckingham, R. C. Gorges, and J. T. Henry, Australian J. Chem., 20, 281 (1967).

[65] M. L. Morris, R. W. Moshier, and R. E. Sievers, Inorg. Chem., 2, 411 (1963).

[66] E. M. Larsen, G. Terry, and J. Leddy, J. Am. Chem. Soc., 75,

[67] R. D. Hill, thesis, University of Manitoba (1962).

[68] B. G. Harvey, H. G. Heal, A. G. Maddock, and E. L. Rowley, J. Chem. Soc., 1947, 1010.

[69]   H. F. Holtzclaw, Jr., J. P. Collman, J. Am. Chem. Soc., 79, 3318
       (1957).

[70]   H. G. Gilman, R. G. Jones, E. Bindschadler, D. Blume, G. Karmas,
       G. A. Martin, Jr., J. F. Nobis, J. R. Thirtle, H. L. Yale, and
       F. A. Yoeman, J. Am. Chem. Soc., 78, 2790 (1956).

[71]   C. Weidenheft, Inorg. Chem., 8, 1174 (1969).

[72]   R. L. Lintvedt, H. D. Russell, and H. F. Holtzclaw, Inorg. Chem.,
       5, 1603 (1966).

[73]   D. Purushotham, V. R. Rao, and Bh. S. V. R. Rao, Anal. Chim.
       Acta, 33, 182 (1965).

[74]   H. Veening, W. E. Bachman, and D. M. Wilkinson, J. Gas Chromatog,
       5, 248 (1967).

[75]   J. A. Bertrand and R. J. Kaplan, Inorg. Chem., 5, 489 (1966).

[76]   R. H. Holm and F. A. Cotton, J. Inorg. Nucl. Chem., 15, 63 (1960).

[77]   Y. Baskin and N. S. Presand, J. Inorg. Nucl. Chem., 25, 1011 (1963).

[78]   E. L. Zebroski, U.S.A.E.C. Rep. BC-63 (1947).

[79]   E. W. Berg and J. T. Truemper, Anal. Chim. Acta, 32, 245 (1965).

[80]   D. C. Luehrs, R. T. Iwamoto, and Kleinberg, J. Inorg. Chem., 4,
       1739 (1965).

[81]   M. L. Cox and R. S. Nyholm, J. Chem. Soc., 1965, 2840.

[82]   R. Sievers, unpublished work.

[83]   J. P. Collman. R. L. Marshall, W. L. Young, III, and S. D. Goldby,
       Inorg. Chem., 1, 704 (1962).

[84]   R. W. Kluiber, Inorg. Chem., 4, 1047 (1965).

[85]   W. R. Walker and N. C. Li, J. Inorg. Nucl. Chem., 27, 2255 (1965).

[86]   R. D. Gillard and G. Wilkinson, J. Chem. Soc., 1963, 5885.

[87]   R. A. Hartman, M. Kilner, and Wojcicki, Inorg. Chem., 6, 34 (1967).

[88]   M. R. Kidd, R. S. Sager, and W. H. Watson, Inorg. Chem., 6, 946
       (1967).

[89]   H. Bauer, J. Blanc, and D. L. Ross, J. Am. Chem. Soc., 86, 5125
       (1964).

[90] R. G. Charles and E. P. Riedel, J. Inorg. Nucl. Chem., 28, 3005
      (1966).

[91] K. Starke, J. Inorg. Nucl. Chem., 25, 823 (1963).

[92] D. E. Grove, N. P. Johnson, C. J. L. Lock, and G. Wilkinson,
      J. Chem. Soc., 1965, 490.

[93] J. P. Collman and M. Yamada, J. Org. Chem., 28, 3017 (1963).

[94] F. Bonati and R. Ugo, J. Organometal. Chem., 11, 341 (1968).

[95] S. C. Chattoraj, A. G. Cupka, and R. E. Sievers, J. Inorg. Nucl.
      Chem., 28, 1937 (1966).

[96] S. Jujiwara, in Spectroscopic Structure of Metal Chelate Compounds
      (K. Nakamoto, ed.), Wiley, New York, 1968, pp. 286-308.

[97] J. Fujita and Y. Shimura in Spectroscopic Structure of Metal Chelate
      Compounds (K. Nakamoto, ed.), Wiley, New York, 1968, pp. 156-215.

[98] L. J. Bellamy and R. F. Branch, J. Chem. Soc., 1954, 4491.

[99] K. Nakamoto, P. J. McCarthy, and A. E. Martell, Nature, 183,
      459 (1959).

[100] R. L. Belford, A. E. Martell, and M. Calvin, J. Inorg. Nucl. Chem.,
      14, 169 (1960).

[101] K. Nakamoto, Y. Morimoto, and A. W. Martell, J. Phys. Chem.,
      66, 346 (1962).

[102] G. A. Domrachev and V. P. Ippolitova, Tr. Khim. Khim. Tekhnol.,
      1966, 227.

[103] C. Y. Liang, Proc. First Int. Conf. Spectrosc., 1967, Vol. 2, p. 302.

[104] L. Wish and R. A. Bolomey, J. Am. Chem. Soc., 72, 4486 (1950).

[105] L. Sacconi and G. Giannoni, J. Chem. Soc., 1954, 2751.

[106] H. I. Feinstein, Microchem. J., 1, 237 (1957).

[107] F. A. Cotton and R. H. Holm, J. Am. Chem. Soc., 82, 2979 (1960).

[108] K. M. Abubacker and M. S. Krishna Brasad, J. Inorg. Nucl. Chem.,
      16, 296 (1961).

[109] J. P. Fackler, Jr., F. A. Cotton, and D. W. Barnum, Inorg. Chem.,
      2, 97 (1963).

[110] K. De Armond and L. S. Forster, Spectrochim. Acta, 19, 1393 (1963).

[111] P. X. Armendarez and L. S. Forster, J. Chem. Phys., 40, 273 (1964).

[112] J. P. Fackler, D. G. Holah, and I. D. Chawla, Proc. Eigth Int. Conf. Coord. Chem., Vienna., 1964, p. 75.

[113] C. H. Ke and N. C. Li, J. Inorg. Nucl. Chem., 28, 2255 (1966).

[114] R. J. Casey and W. R. Walker, J. Inorg. Nucl. Chem., 29, 1301 (1967).

[115] L. L. Funck and T. R. Ortolano, Inorg. Chem., 7, 567 (1968).

[116] A. C. Adams and E. M. Larsen, J. Am. Chem. Soc., 85, 3508 (1963).

[117] R. L. Scruggs, P. Kim, and N. C. Li, J. Phys. Chem., 67, 2194 (1963).

[118] R. A. Palmer, R. C. Fay, and T. S. Piper, Inorg. Chem., 3, 875 (1964).

[119] R. C. Fay, and T. S. Piper, Inorg. Chem., 3, 348 (1964).

[120] A. C. Adams and E. M. Larsen, Inorg. Chem., 5, 228 (1966).

[121] A. C. Adams and E. M. Larsen, Inorg. Chem., 5, 814 (1966).

[122] W. R. Walker and N. C. Li, J. Inorg. Nucl. Chem., 27, 411 (1965).

[123] M. M. McGrady and R. S. Tobias, J. Am. Chem. Soc., 87, 1909 (1965).

[124] R. G. Linck and R. E. Sievers, Inorg. Chem., 5, 806 (1966).

[125] D. Gibson, J. Lewis, and C. Oldham, J. Chem. Soc., 1966, 1453.

[126] J. J. Fortman and R. E. Sievers, Inorg. Chem., 6, 2022 (1967).

[127] R. C. Fay and N. Serpone, J. Am. Chem. Soc., 90, 5701 (1968).

[128] N. Filipescu, M. R. Kagan, and F. A. Serafin, Nature, 196, 467 (1962).

[129] N. Filipescu, W. F. Sager, and F. A. Serefin, J. Phys. Chem., 68, 3324 (1964).

[130] M. L. Bhaumik, L. J. Nugent, and L. Ferder, J. Chem. Phys., 41, 1158 (1964).

[131] M. L. Bhaumik, J. Chem. Phys., 40, 3711 (1964).

[132] N. McAvoy, N. Filipescu, M. R. Kagan, and F. A. Serafin, J. Phys. Chem. Solids, 25, 461 (1964).

[133] M. A. El-Sayed and M. L. Bhaumik, J. Chem. Phys., 39, 2391 (1963).

[134] M. L. Bhaumik, J. Inorg. Nucl. Chem., 27, 261 (1965).

[135] M. L. Bhaumik and M. A. El-Sayed, J. Phys. Chem., 69, 275 (1965).

[136] M. L. Bhaumik and M. A. El-Sayed, J. Chem. Phys., 42, 787 (1965).

[137] M. L. Bhaumik, J. Inorg. Nucl. Chem., 27, 243 (1965).

[138] M. L. Bhaumik and N. J. Nugent, J. Chem. Phys., 43, 1680 (1965).

[139] E. V. Melent'eva, L. I. Kononenko, E. G. Koltunova, and N. S. Plouektov, Zh. Prikl. Spektrosk., 5, 328 (1966).

[140] V. V. Kuznetosova, A. N. Sevchenko and V. S. Khomenko., Zh. Prikl. Spektrosk., 5, 480 (1966).

[141] H. Samelson, C. Brecher, and A. Lempicki, J. Mol. Spectro, 19, 349 (1966).

[142] N. Filipescu, G. W. Mushrush, C. R. Hurt, and N. McAvoy, Nature, 211, 960 (1966).

[143] T. M. Shepherd, Nature, 211, 745 (1966).

[144] W. R. Dawson, J. L. Kropp, and M. W. Windsor, J. Chem. Phys., 45, 2410 (1966).

[145] C. Reichert, J. B. Westmore, and H. D. Gesser, Chem. Commun., 1967, 782.

[146] G. M. Bancroft, C. Reichert, J. B. Westmore, and H. D. Gesser, Inorg. Chem., 8, 474 (1969).

[147] E. Sheperd and W. W. Meinke, U.S. Atomic Energy Comm. AECU-3879, (1958).

[148] A. V. Lavrukhina, V. Kourzhim, and L. V. Filatova, Radiokhimiya, 1, 204 (1959).

[149] F. J. Hagemann, U.S. Pat. 2,632,763 (1953).

[150] V. M. Dziomko, Tr. Vses. Nauchn.-Issled. Inst. Khim. Reaktivov, 25, 183 (1963).

[151] H. C. Eshelman, J. A. Dean, O. Menis and T. C. Rains, Anal. Chem.,
       31, 183 (1959).

[152] D. Dyrssen, M. Heffez, and T. Sekine, J. Inorg. Nucl. Chem., 16,
       367 (1961).

[153] T. Sekine and D. Dyrssen, Talanta, 11, 867 (1964).

[154] T. Sekine and D. Dyrssen, J. Inorg. Nucl. Chem., 29, 1481 (1967).

[155] L. B. Werner, I. Perlman, and M. Calvin, U.S. Pat. 2,894,805
       (1959).

[156] T. Sekine, Acta Chem. Scand., 19, 1435 (1965).

[157] T. Sekine, Acta Chem. Scand., 19, 1469 (1965).

[158] J. Stary, AEC Accession No. 22093, Rep. No. JINR-P-2,000, 1965.

[159] J. Stary, Radiokhimiya, 8, 504 (1966).

[160] J. Stary, Sb. Ref. Celostatni Radiochem. Konf., 3, 7 (1964).

[161] J. Sekine, M. Sakairi, and Y. Hasegawa, Bull. Chem. Soc. Japan,
       39, 2141 (1966).

[162] I. Akaza, Bull. Chem. Soc. Japan, 39, 971 (1966).

[163] T. Sekine and M. Sakairi, Bull. Chem. Soc. Japan, 40, 261 (1967).

[164] F. L. Moore, Anal. Chem., 38, 1872 (1966).

[165] T. Sekine and D. Dyrssen, Anal. Chim. Acta, 37, 217 (1967).

[166] I. P. Alimarin and Y. A. Zolotov, Talanta, 9, 891 (1962).

[167] E. Uhlemann and H. Mueller, Z. Chem., 8, 185 (1968).

[168] G. K. Schweitzer and L. H. Howe, III, Anal. Chim. Acta, 37, 316
       (1967).

[169] Y. A. Zolotov, Dokl. Akad. Nauk. SSSR, 162, 577 (1965).

[170] A. K. De and S. K. Majumdar, Anal. Chim. Acta, 27, 153 (1962).

[171] H. Irving, Proc. Symp. Coord. Chem., 1964, 219.

[172] O. M. Petrukhin, L. A. Izosenkova, I. N. Marov, N. Dubrov and
       Y. A. Zoltov, Zh. Neorg. Khim., 12, 1407 (1967).

[173] P. Crowther and F. L. Moore, Anal. Chem., 35, 2081 (1963).

[174] M. Kyrs and S. Podesva, Czech., 111,376 (1964).

[175] D. Dyrssen, Proc. Symp. Coord. Chem., 1964, 231.

[176] R. J. Casey, J. J. M. Fardy, and W. R. Walker, J. Inorg. Nucl. Chem., 29, 1139 (1967).

[177] L. Newman and P. Klotz, AEC Accession No. 38855, Rep. No. BNL-10350, 1966.

[178] I. J. Gal and R. M. Nikolic, J. Inorg. Nucl. Chem., 28, 563 (1966).

[179] S. M. Khopkar and A. K. De, Z. Anal. Chem., 171, 241 (1959).

[180] T. Sekine and D. Dyrssen, J. Inorg. Nucl. Chem., 26, 1727 (1964).

[181] T. Sekine and D. Dyrssen, J. Inorg. Nucl. Chem., 26, 2013 (1964).

[182] S. M. Khopkar and A. K. De, Anal. Chim. Acta, 22, 223 (1960).

[183] C. Testa, Anal. Chim. Acta, 25, 525 (1961).

[184] F. L. Moore, W. D. Fairman, J. G. Ganchoff, and J. G. Surak, Anal. Chem., 31, 1148 (1959).

[185] H. L. Finston and Y. Inoue, J. Inorg. Nucl. Chem., 29, 199 (1967).

[186] H. Onishi and Y. Toita, Bunseki Kagaku, 14, 462 (1965).

[187] R. M. Nikolic and I. T. Gal, Croat. Chem. Acta, 38, 17 (1966).

[188] E. M. Larsen and G. Terry, J. Am. Chem. Soc., 75, 1560 (1953).

[189] R. G. Deshpande, P. K. Khopkar, C. L. Rao, and H. D. Sharma, J. Inorg. Nucl. Chem., 27, 2171 (1965).

[190] L. P. Varga and D. N. Hume, Inorg. Chem., 2, 201 (1963).

[191] V. M. Peshkova and An P'Eng, Vestn. Mosk. Univ. Ser. II: Khim, 18, 40 (1963).

[192] V. M. Peshkova and An P'Eng, Zh. Neorg. Khim., 7, 1484 (1962).

[193] E. H. Huffman, G. N. Iddings, R. M. Osborne, and G. V. Shalimoff, J. Am. Chem. Soc., 77, 881 (1955).

[194] J. Hala, J. Inorg. Nucl. Chem., 29, 187 (1967).

[195] T. Sekine and D. Dyrssen, J. Inorg. Nucl. Chem., 29, 1489 (1967).

[196] M. Pivonkova and M. Kyrs, Chem. Listy, 6, 20 (1967).

[197] H. Akaiwa and H. Kawamoto, Bunseki Kagaku, 16, 359 (1967).

[198] A. K. De and M. S. Rahaman, Anal. Chem., 35, 159 (1963).

[199] H. Yoshida, H. Nagai, and H. Onishi, Talanta, 13, 37 (1966).

[200] D. C. Perricos and J. A. Thomassen, Kjeller Rept., KR-83 (1964).

[201] A. K. De and M. S. Rahaman, Anal. Chem., 36, 685 (1964).

[202] A. Jurriaanse and F. L. Moore, Anal. Chem., 39, 494 (1967).

[203] H. L. Scherff and G. Herrmann, J. Inorg. Nucl. Chem., 11, 247 (1959).

[204] C. J. Hardy and D. Scargill, J. Inorg. Nucl. Chem., 9, 322 (1959).

[205] A. K. De and M. S. Rahaman, Anal. Chim. Acta, 27, 591 (1962).

[206] A. Chesné , Energie Nucléaire, 1, 210 (1959).

[207] M. Calvin, U.S. Pat. 2,856,418 (1958).

[208] R. A. Schneider, Anal. Chem., 34, 522 (1962).

[209] Y. A. Zolotov and I. P. Alimarin, J. Inorg. Nucl. Chem., 25, 691 (1963).

[210] I. V. Shilin and V. K. Nazarov, Radiokhimiya, 8, 514 (1966).

[211] L. B. Magnusson, U.S. Pat. 2,830,066 (1958).

[212] Y. A. Zolotov and I. P. Alimarin, Radiokhimiya, 4, 272 (1962).

[213] G. Bouissieres and J. Vernois, Compt. Rend., 244, 2508 (1957).

[214] Q. Van Winkle, U.S. Pat. 2,895,791 (1959).

[215] B. Myasoedov and R. Muxart, Bull. Soc. Chim. France, 1962, 237.

[216] B. Myasoedov and R. Muxart, Zh. Analit. Khim., 17, 340 (1962).

[217] E. S. Pal'shin and B. Myashoedov, Zh. Analit. Khim, 18, 750 (1963).

[218] F. Boulanger and R. Guillaumont, Bull. Soc. Chim. France, 1964, 3031.

[219] R. Guillaumont, Comp. Rend., 260, 1416 (1965).

[220] R. Guillaumont and G. Bouissieres, Bull. Soc. Chim. France, 1964, 2098.

[221] A. V. Davydov, B. Myasoedov, Y. P. Novikov, P. N. Palei, and E. S. Pal'shin, Tr. Komis. Po. Analit. Khim. Akad. Nauk. SSSR, 15, 64 (1965).

[222] R. Guillaumont, Bull. Soc. Chim. France, 1965, 2106.

[223] R. Guillaumont, Colloq. Intern. Centre Natl. Rech. Sci., 154, 165 (1966).

[224] B. T. Kolarich, V. A. Ryan, and R. P. Schuman, J. Inorg. Nucl. Chem., 29, 783 (1967).

[225] A. K. De and M. S. Rahaman, Analyst, 89, 795 (1964).

[226] J. G. Cunninghame and G. L. Miles, J. Appl. Chem., 7, 72 (1957).

[227] D. H. W. den Boer and Z. I. Dizdar, JENER Rept. No. 45, 1965.

[228] A. S. G. Mazumdar and C. K. Sivaramakrishnan, J. Inorg. Nucl.
       Chem., 27, 2423 (1965).

[229] O. Cristallini and M. Rudelli, Conf. Interam. Radioquim., Montevideo,
       1963, 141.

[230] J. G. Cunninghame and G. L. Miles, J. Inorg. Nucl. Chem., 3, 54
       (1956).

[231] H. Irving and D. N. Edgington, Chem. Ind. (London), 1961, 77.

[232] H. W. Crandall, J. R. Thomas, and J. C. Reid, U.S. Atomic Energy
       Comm., CN-2657 (1957).

[233] T. R. Hicks and H. W. Crandall, U.S. Atomic Energy Comm.,
       UCRL-912 (1957).

[234] A. V. Rangnekar and S. M. Khopkar, Bull. Chem. Soc. Japan, 39,
       2169 (1966).

[235] S. Oki, T. Omori, T. Wakahayashi, and N. Suzuki, J. Inorg. Nucl.
       Chem., 27, 1141 (1965).

[236] T. Wakahayashi, S. Oki, T. Omori, and N. Suzuki, J. Inorg. Nucl.
       Chem., 26, 2255 (1964).

[237] M. P. Belopolskii and N. P. Popov, Tr. Vses. Nauchn.-Issled.
       Geol. Inst., 117, 111 (1964).

[238] T. Sekine, A. Koizumi, and M. Sakairi, Bull. Chem. Soc. Japan,
       39, 2681 (1966).

[239] T. Shigematsu, M. Tabushi, M. Matsui, Y. Nishikawa, and S. Goda,
       Nippon Kogaku Zasshi, 84, 263 (1963).

[240] T. Takenchi, M. Suzuki, and M. Yanagisawa, Anal. Chim. Acta, 36,
       258 (1966).

[241] G. K. Schweitzer and W. V. Willis, Anal. Chim. Acta, 36, 77 (1966).

[242] J. R. Stokely and F. L. Moore, Anal. Chem., 36, 1203 (1964).

[243] Z. Kolarik and H. Pankova, Collection Czech. Chem. Commun.,
       27, 166 (1962).

[244] W. C. Johnson, Jr., Anal. Chem., 38, 954 (1966).

[245] M. Barrachina and R. Sauvagnac, Comm. Energie At., Rappt. No.
       C.E.A. 2166 (1962).

[246] V. V. Bagreev and Y. A. Zolotov, Zh. Analit. Khim., 18, 425 (1963).

[247] N. Suzuki and T. Kato, Sci. Rept. Tohoku Univ., 43, 152 (1959).

[248] V. I. Spitsyn, A. F. Kuzina, N. N. Zamoshnikova, and T. S. Tagil,
       Dokl. Akad. Nauk. SSSR, 144, 1066 (1962).

[249] R. F. Bogucki, Y. Murakami, and A. E. Martell, J. Am. Chem. Soc.,
       82, 5608 (1960).

[250] R. Collee, Rev. Universelle Mines, 21, 330 (1965).

[251] W. W. Meinke and R. E. Anderson, Anal. Chem., 24, 708 (1952).

[252] A. Liberti, V. Chiantella, and F. Corigliano, J. Inorg. Nucl. Chem.,
       25, 415 (1963).

[253] A. Liberti, V. Chiantella, and F. Corigliano, Ann. Chim., 52, 813
       (1962).

[254] A. K. De and M. S. Rahaman, Anal. Chim. Acta, 31, 81 (1964).

[255] G. K. Schweitzer and A. D. Norton, Anal. Chim. Acta, 30, 119 (1964).

[256] V. V. Bagreev and Y. A. Zolotov, Zh. Analit. Khim., 17, 852 (1962).

[257] T. Shigematsu, M. Tabushi, and M. Matsui, Bull. Chem. Soc. Japan,
       37, 1333 (1964).

[258] S. Takei, Nippon Kagaku Zasshi, 87, 949 (1966).

[259] R. A. Day, Jr., and R. M. Powers, J. Am. Chem. Soc., 76, 3895
       (1954).

[260] G. N. Walton, F. Barker and G. Byfleet, At. Energy Res. Estab.,
       No. C/R768 (1955).

[261] S. M. Khopkar and A. K. De, Chem. Ind. (London), 1959, 291.

[262] E. L. King, U.S. Atomic Energy Comm., TID-5290, 269 (1958).

[263] M. A. Awwal, Pakistan J. Sci. Ind. Res., 8, 70 (1965).

[264] N. Milich, O. M. Petrukhin, and Y. A. Zolotov, Zh. Neorg. Khim.,
       9, 2664 (1964).

[265] K. Batzar, D. E. Goldberg and L. Newman, J. Inorg. Nucl. Chem.,
       29, 1511 (1967).

[266] H. Yoshida, Bull. Chem. Soc. Japan, 39, 1810 (1966).

[267] A. Ikehata and T. Shimizu, Bull. Chem. Soc. Japan, 38, 1385 (1965).

[268] G. K. Schweitzer and S. W. McCarty, Anal. Chim. Acta, 29, 56 (1963).

[269] W. R. Walker and M. S. Farrell, J. Inorg. Nucl. Chem., 28, 1483 (1966).

[270] M. Kyrs, J. Rais, and P. Pistek, Sb. Ref. Celostatni Radiochem. Konf., 3, 5 (1964).

[271] B. Z. Iofa and A. S. Yushchenko, Zh. Neorg. Khim., 10, 558 (1965).

[272] H. W. Crandall and J. R. Thomas, U.S. Pat. 2,892,681 (1959).

[273] E. H. Huffman and G. M. Iddings, U.S. Atomic Energy Comm., UCRL-377 (1949).

[274] G. P. Anetova, N. V. Mel'chakova, and V. M. Peshkova, Vestn. Mosk. Univ. Khim., 23, 102 (1968).

[275] S. F. March, W. J. Maeck, G. L. Booman, and J. E. Rein, Anal. Chem., 33, 870 (1961).

[276] H. Onishi and C. V. Banks, Anal. Chem., 35, 1887 (1963).

[277] H. Yoshida, H. Nagai and H. Onishi, Bunseki Kagaku, 15, 513 (1966).

[278] M. A. Awwal, Anal. Chem., 35, 2048 (1963).

[279] M. Khopkar and A. K. De, Anal. Chem., 32, 478 (1960).

[280] A. K. Lavrukhina, L. V. Yukina, and Z. V. Khromchenko, Tr. Komis. Po Analit. Khim., Akad. Nauk. SSSR, 14, 202 (1963).

[281] V. T. Mishchenko, R. S. Lauer, N. P. Efryushina, and N. S. Poluektov, Zh. Analit. Khim., 20, 1073 (1965).

[282] P. G. Manning, Can. J. Chem., 44, 3057 (1966).

[283] P. G. Manning and C. B. Monk, Trans. Faraday Soc., 58, 938 (1962).

[284] T. Sekine and M. Ono, Bull. Chem. Soc. Japan, 38, 2087 (1965).

[285] F. W. Cornish, At. Energy Res. Estab., T/M-145 (1957).

[286] E. V. Melent'eva, N. S. Poluektov, and L. I. Kononenko, Zh. Anal. Khim., 22, 187 (1967).

[287] G. V. Korpusov, G. V. Tsygankova, and E. G. Goryacheva, Tsvetn. Metal., 37, 63 (1967).

[288] L. Genov, G. Kasabov, and I. Cholakova, Monatsh. Chem., 96, 2005 (1965).

[289] T. K. Keenan and J. F. Suttle, J. Am. Chem. Soc., 76, 2184 (1954).

[290] W. G. Scribner, W. J. Treat, J. D. Weis, and R. W. Moshier, Anal. Chem., 37, 1136 (1965).

[291] W. G. Scribner, M. J. Borchers, and W. J. Treat, Anal. Chem., 38, 1779 (1966).

[292] W. G. Scribner and A. M. Kotecki, Anal. Chem., 37, 1304 (1965).

[293] B. G. Schultz and E. M. Larsen, J. Am. Chem. Soc., 72, 3610 (1950).

[294] N. V. Mel'chakova, N. N. Mogdesieva, Y. K. Yur'ev, and V. M. Peshkova, Vestn. Mosk. Univ. Khim., 21, 82 (1966).

[295] V. M. Peshkova, I. P. Efimov, and N. N. Magdesieva, Zh. Analit. Khim., 21, 499 (1966).

[296] U.S. Atomic Energy Commission, Brit. Pat. 895,676 (1962).

[297] W. D. Ross and R. E. Sievers, Talanta, 5, 87 (1968).

[298] R. S. Juvet and R. P. Durbin, J. Gas Chromatog., 1, 14 (1963).

[299] R. D. Hill and H. Gesser, J. Gas Chromatog., 1, 11 (1963).

[300] W. D. Ross and G. Wheeler, Anal. Chem., 36, 266 (1964).

[301] W. D. Ross, Anal. Chem., 35, 1597 (1963).

[302] D. K. Albert, Anal. Chem., 36, 2034 (1964).

[303] W. D. Ross, R. E. Sievers, and G. Wheeler, Jr., Anal. Chem., 37, 598 (1965).

[304] R. E. Sievers, J. W. Connolly, and W. D. Ross, J. Gas Chromatog., 5, 241 (1967).

[305] R. E. Sievers, R. W. Moshier, and M. L. Morris, Inorg. Chem., 1, 966 (1962).

[306] M. Tanaka, T. Shono, and K. Shinra, Anal. Chim. Acta, 43, 157 (1968).

[307] G. P. Morie and T. R. Sweet, Anal. Chem., 37, 1552 (1965).

[308] R. E. Sievers, K. J. Eisentraut, C. E. Springer, Jr., and D. W. Meek, Adv. Chem., 71, 141 (1967).

[309] H. Veening and K. Huber, J. Gas Chromatog., 6, 326 (1968).

[310] T. Shigematsu, M. Matsui, and K. Utsunomiya, Bull. Chem. Soc. Japan, 41, 763 (1968).

[311] K. Tanikawa, K. Hirano, and K. Arakawa, J. Chem. Pharm. Bull., 15, 915 (1967).

[312] R. S. Juvet and F. Zado, Advances in Chromatography, Vol. 1, Dekker, New York, 1966.

[313] R. E. Sievers, 16th Ann. Summer Sym. Anal. Chem., Tucson, Ariz., June 19, 1963.

[314] J. E. Schwarberg, R. W. Moshier, and J. H. Walsh, Talanta, 11, 1213 (1964).

[315] R. W. Moshier and J. E. Schwarberg, Talanta, 13, 445 (1966).

[316] C. Genty, C. Houin, and R. Schott, 7th Intern. Symp. Gas Chromatog. Copenhagen, June, 1968.

[317] R. E. Sievers, B. W. Ponder, and R. W. Moshier, 141st Am. Chem. Soc. Mtg., Washington, D. C., March, 1962; Anon., Chem. Eng. News, 40, 50 (1962).

[318] B. Feibush and E. Gil-Av, in Advances in Chromatography, 1967 (A. Zlatkis, ed.), Preston Tech. Abstracts Co., Evanston, Ill., p. 100; and references therein.

[319] R. E. Sievers, in Gas Chromatography 1966 (A. B. Littlewood, ed.), Elsevier, Amsterdam, 1966, p. 292.

[320] R. E. Sievers, J. W. Connolly, A. S. Hilton, M. F. Richardson, J. E. Schwarberg, and W. D. Ross, 155th Am. Chem. Soc. Mtg., San Francisco, Calif. April, 1968.

[321] K. J. Eisentraut and R. E. Sievers, J. Inorg. Nucl. Chem., 29, 1931 (1967).

[322] J. E. Schwarberg, D. R. Gere, R. E. Sievers, and K. J. Eisentraut, Inorg. Chem., 6, 1933 (1967); and references therein.

[323] J. E. Sicre, J. T. Dubois, K. J. Eisentraut, and R. E. Sievers, Proc. 10th Intern. Conf. Coord. Chem., Chem. Soc. Japan, Tokyo, 1967.

[324] G. P. Morie and T. R. Sweet, Anal. Chim. Acta, 34, 314 (1966).

[325] W. D. Ross and R. E. Sievers, Sixth Intern. Symp. Gas Chromatog.,
Rome, Italy, September, 1966.

[326] W. D. Ross and R. E. Sievers, Anal. Chem., 41, 1109 (1969).

[327] W. Mertz, Physiol. Rev., 49, 163 (1969).

[328] J. Savory, P. Mushak, N. O. Roszel, and F. W. Sunderman, Jr.,
Federation Proc., 27 (3145), 777, (1968).

[329] J. Savory, P. Mushak, and F. W. Sunderman, Jr., J. Chromatog.
Sci., 7, 674 (1969).

[330] J. Savory, P. Mushak, F. W. Sunderman, Jr., R. H. Estes and
N. O. Roszel, Anal. Chem., 42, 294 (1970).

[331] L. C. Hansen, M.S. dissertation, University of Cincinnati, 1970.

[332] J. Savory, M. Glenn, P. Mushak, and J. A. Ahlstrom, unpublished
results.

[333] M. L. Taylor, E. L. Arnold, and R. E. Sievers, Anal. Letters, 1,
735 (1968).

[334] M. H. Noweir and J. Cholack, Environ. Sci. Technol., 3, 927 (1969).

Chapter 3

NMR SPECTRA AND CHARACTERISTIC

FREQUENCIES OF COMPOUNDS CONTAINING N-S-F BONDS

Hans-Georg Horn

Lehrstuhl für Anorganische Chemie II
Ruhr-Universität Bochum
German Federal Republic

I. INTRODUCTION

Compounds which contain a N-S-F bond are not only of academic interest but have practical uses as well. Those in which perfluorocarbon groups have been introduced have been tested and found useful as fungicides and herbicides. Since the discovery of NSF$_3$ by Glemser [1] and his co-workers as early as 1956, many of these interesting compounds have been synthesized. The chemistry and physical properties of these substances have been widely discussed [1, 17], but a compilation of NMR data of this class of sulfur compounds has not been published, although data on sulfur fluorine compounds (and a few N-S-F compounds), ^{19}F shifting values, and coupling constants are mentioned in extensive tabulations [47]. Another paper [42] gives interpretations of NMR absorptions of SF$_5$ compounds, but compounds with NSF$_5$ groups were not known in 1956. Therefore, it is the aim of this article to collect all data spread throughout the literature and to try to find out whether there are correlations between special types of groups and the shifts and intervals for NMR absorption frequencies. At the same time attempts have

been made to find intervals for characteristic frequencies in the ir spectra of
the compounds under discussion.

A. Chemical Shift

Because of the diamagnetic shielding of the nucleus by electrons of an
atom, an outer magnetic field is somewhat altered near the nucleus. Since
this diamagnetic shielding is a function of the electron density and the distance of the electrons from the nucleus, and these factors depend on the
chemical surrounding of the nucleus in question, which means the chemical
bonding of the atom, one can find differences in the resonance frequency for
the same kind of atoms in different compounds. This effect is called the
chemical shift. The chemical shift is the same for all atoms of a molecule
which are chemically equivalent, and can be assumed to be a measure of the
electronic shielding of the atom in question. NMR measurements, therefore,
do not have any absolute character and a reference standard is needed.
Numerous substances have been used as internal and external standards in
^{19}F-NMR spectroscopy. It seems of general interest to tabulate some of
these compounds (see Table 1). Since most of the chemical shifts have been
determined relative to $CFCl_3$ (Freon 11) as an internal or external standard,
only a few values had to be converted to this scale.

TABLE 1

Reference Compounds in ^{19}F-NMR

$CFCl_3$ (Freon 11)	0.0	F_2	-430.5
$CCl_2F\ CCl_2F$ (Freon 112)	65.1	KF/H_2O	117.7
CF_3COOH	76.6	BF_3	125.0
C_4F_8	138.1	$BF_3 \cdot O(C_2H_5)_2$	160.7

On this scale nearly all of the values of the chemical shift have a negative sign (see Tables 3-11); in general the most negative values are found for compounds with sulfur of oxidation number 6, while fluorine bonded to tetravalent sulfur gives rise to positive values as well. This means that, generally, the fluorine in these substances is electronically less shielded than in the reference compound, $CFCl_3$. There is one extreme exception from this rule: In $N \equiv S - F$, which has been synthesized by Glemser and his co-workers, the fluorine atom bonded to tetravalent sulfur is less shielded than in any other compound containing a nitrogen-sulfur-fluorine bond. It may be noted that $N \equiv S - F$ has a shift of -240. 3 ppm relative to $CFCl_3$, and NSF_3, in addition to a change of the oxidation number of sulfur, has nearly a "one-third" shift of -66. 8 ppm compared to NSF. This strong downfield shift of NSF may be due to the triple bond between nitrogen and sulfur, where the electrons may be attracted more strongly by N than by S, thus effecting a withdrawal of electrons from fluorine. One may compare the electronegativity of elemental nitrogen (3. 0) and sulfur (2. 5). The electronegativity of fluorine (4. 0) would not seem to be that influential, as fluorine is the first member of the halogen family and it is well known that in many cases, the first member of such a series shows deviations from the physical property in question. (See, for example, the discussion in Sec. I. C.). Furthermore, the fluorine may become π-bonded, which means that it is less shielded than the fluorine in $CFCl_3$. The large negative shift of fluorine atoms bonded to hexavalent sulfur may be explained in a similar way. With an increasing oxidation number, the nuclear charge of sulfur increases, thus attracting the electrons from fluorine, and leaving the fluorine atom less shielded compared with tetravalent sulfur compounds.

It is of interest that the N-S bond distance [73] is of the same order of magnitude in NSF_3 (r=1. 416 Å) and NSF (r=1. 446 Å) while the valence force constants (NSF_3:12. 55, NSF:10. 71) are also close together.

It was not possible to indicate in the tables whether the $CFCl_3$ has been used as an external or internal reference, although it seems that for most of the measurements $CFCl_3$ was an external standard. The ^{1}H-NMR absorption measurements were always done in presence of tetramethylsilane $(CH_3)_4Si$ as a reference compound. In general, numbers following a formula or "compound" indicate the number of the compound in the tables; bracketed numbers indicate references.

B. Coupling Constants

As could be assumed from the above definition of the chemical shift, a compound, for example $CF_2(NSF_2)_2$, should give rise to only two ^{19}F-NMR signals of intensity 1:2 because of the presence of two kinds of chemically nonequivalent fluorine atoms. But a quintuplet appears for the fluorine atoms attached to carbon and a triplet is found for the fluorine atoms attached to sulfur. This splitting of the "normal" signal arises from spin-spin coupling, caused by magnetic polarization of two nuclei by means of the bonding electrons between the two nuclei in question.

First-order spectra of this type obey simple rules. The interaction between two nuclei is explained through use of the coupling constant, which is a valuable tool for determining the structure of unknown molecules. Such a fine structure of a ^{19}F-NMR spectrum is independent of temperature and the applied magnetic field. The lines of such a multiplet structure are equally spaced, with a separation equal to the coupling constant. A monotonical decrease of the magnitude of the coupling constant is found in ^{1}H-NMR spectroscopy with an increase of the number of bonds between interacting nuclei. This phenomenon can be seen from the tables in some cases. Fluorine shows another behavior; it is often found that coupling constants between nearer nuclei are smaller than those that are found when separation occurs by, say, three or more bonds. Several factors may account for these events. Coupling may be produced by bonding electrons as well as by nonbonding π electrons. Furthermore, space coupling may be responsible for this effect, although no theoretical evaluation has been published nor any ideas of the mechanism proposed. But any other kind of mechanism can hardly be imagined for spin-spin interactions over five or six σ bonds (long-range coupling) as has already been observed. Nothing can be said about the sign of these coupling constants.

Normally no fine structure appears which involves the coupling of nitrogen and fluorine. This fact is due to the electric quadrupole moment of nitrogen, which causes relaxation broadening. Thus in most cases the signal becomes so broad that it is not separable from the background, although sometimes a broad signal appears at room temperature which shows splitting on cooling. This is especially the case in molecules with N-F bonds; for instance, in compound 36, F-N-S=O, F-N coupling gives rise to a triplet with a coupling constant of 112 Hz. The other triplet structures in the ^{19}F-NMR spectra of compounds containing NF_2 or NF groups where coupling constants vary around 100 Hz, are discussed in connection with molecules of the special section. Thiazyl trifluoride, $N\equiv SF_3$ (134, 134a) shows N-F spin-spin coupling. This is the only example for N-F coupling over two bonds. As expected, the coupling constant of 26 Hz is much lower than for compounds with N-F bonds, and the triplet signal is not as sharp as in molecules with carbon fluorine bonds. $N\equiv SF$ (31) does not show any fine structure.

C. Characteristic Frequencies

Often polyatomic molecules give rise to complex vibrational spectra because of their low symmetry, mechanical coupling of vibrations, etc. Thus it is difficult to assign all bands of the observed spectra to motions of every atom of the molecule. This means that the bands of an infrared spectrum may be assigned to the separated vibrations of two atoms of a more complex molecule of low symmetry. Characteristic bands are, e.g., N-H, C-H, $-C\equiv O$ (in complexes), P=O, $\overset{\ominus}{N}=\overset{\oplus}{N}=\overset{\ominus}{N}$, etc. In N-S-F chemistry, characteristic frequencies are the N=S, S=O, and S-F vibrations. The first is found in the range of 1150–1450 cm^{-1}, the second in the same region, the latter in the interval 650–920 cm^{-1}. All these bands may appear from medium to very

strong intensity. Sometimes complications occur because the N-F and C-F
vibrations may overlap those of N=S, S=O, and S-F. In these cases the
intensity and contour of the bands often allow the assignment of these absorptions to characteristic vibrations. The valence force constant f_{NS} has been
calculated to be 9.89 mdyn/$\overset{\circ}{A}$ (N_{NS}=2.2) by Muller et al. [74] on the assump-
tion of a two-mass model for the compound $F_5S-N=SF_2$. In this case, ν_{NS}
seems to be very characteristic. Since for a variety of the other N=S
compounds, ν_{NS} has been found in the same region, the differences of ν_{NS}
are explained by differences of bonding and mechanical coupling. This leads
to the idea that all compounds of the type R-NSF$_2$ have a bond order higher
than 2 and their formulas may be represented by the canonical structures

$$R-\bar{N}=\bar{S}F_2 \Longleftrightarrow R-\overset{\oplus}{N}\equiv\overset{\ominus}{S}F_2$$
$$\text{I} \qquad\qquad\qquad \text{II}$$

Similar calculations of valence force constants and bond orders in
NSF$_3$, NSF, and other compounds containing N-S bonds by Glemser et al.
[73] gave N_{NS}=2.7 (NSF$_3$) and N_{NS}=2.4 (NSF), which indicates that the
above considerations and structures in general seem to be correct and
explain the high wavenumbers of ν_{NS} of molecules of the RNSF$_2$ type. While
the assignments of ν_{NS} and ν_{SO} given in Tables 3-10 are tentative, in one
case a series of molecules HalNSO (Hal=F, Cl, Br, I) have been investigated
more thoroughly [84] by both infrared and Raman spectra [gaseous (ir) and
liquid (Ra) state, INSO solid]. The region of the vibrations ν_{NS}(cm^{-1}) and
ν_{SO}(cm^{-1}) for compounds HalNSO may be compared with the assigned
ν_{NS}(cm^{-1}) in the series of similar compounds, HalNSF$_2$ (Hal=F, Cl, Br, I)
[6, 16]. Electronegativities of the halogens are given in Table 2.

Infrared frequencies not shown in Table 2 and Raman spectra of HalNSO
agree well in spite of the different aggregate states. The vibration ν_{NS} for
the series HalNSO and HalNSF$_2$ shows differences around 200 wavenumbers,
which may be due to the electronegativity differences of O and F and the
above-mentioned structure II which contributes to the hybrid structure.
If ν_{NS} is plotted against the electronegativity of the halogen atoms of HalNSO
or HalNSF$_2$, one should expect a straight line. This was shown by Glemser

TABLE 2

Vibrations of HalNSO Compounds

Atom (Hal)	Electronegativity	Compound	Raman ν_{NS}	ν_{SO}	Compound	Infrared ν_{NS}
F	4.0	FNSO	995	1230	$FNSF_2$	1150
Cl	3.0	ClNSO	989	1221	$ClNSF_2$	1200
Br	2.8	BrNSO	1000	1214	$BrNSF_2$	1215
I	2.5	INSO	1028	1277	$INSF_2$	1235

et al. [16] for the compounds $HalNSF_2$ (Hal=Cl, Br, I). But the data on $FNSF_2$ published shortly afterwards by the same author does not fit a straight line. The same result has been obtained for the series HalNSO (ν_{NS} versus electronegativity or ν_{SO} versus electronegativity); In each case the first member of the halogen family does not follow a straight-line relationship. The reason for this may be p_π-d_π interactions between fluorine and sulfur.

Several techniques have been applied for obtaining the spectra: Spectra of gaseous and solid compounds have been taken in the usual way, while liquid substances were used as capillary films.

Abbreviations in the tables have the usual meanings. Question marks without parentheses after wavenumbers mean that these wavenumbers have been taken from published data of the literature cited and are of doubtful character.

II. COMPOUNDS CONTAINING NSF_2 GROUPS

A. NMR Data

In the first column of Table 3, the shift δ_{S-F} (ppm) of the NSF_2 group is given. The range in which NMR absorption occurs for these molecules varies from −16.7 ppm for $FNSF_2$ to −66.3 ppm for $C_6H_5NSF_2$ (26). Again the fluorine compound of the $RNSF_2$ series (R=F) shows an abnormal behavior, while the next most negative shift of −31.6 ppm has been determined for $(CF_3)_2C=NNSF_2$ (17), so that the range at which absorption does occur is from −30 to −70 ppm. Shifts of the other groups connected to the NSF_2 group are given in Table 4.

No general rules can be given for these shifts, because they depend on the different kind of groups and the distance of groups attached to NSF_2. Trifluoromethyl groups show a shift of between +50 and +80 ppm while the CF_2 group gives rise to peaks in the +40 ppm region. There is one exception, $CF_3CF_2NSF_2$ (15), which shows a signal at +163.4 ppm for $\delta (CF_2)$. Spin-spin coupling is often observed between the NFS_2 group and other parts of the molecule; therefore, the multiplicity is given in parentheses after the chemical shift. Thus, for instance, coupling between fluorine bonded directly to nitrogen and the fluorine of the NSF_2 group in (2) is found to be a triplet spaced by 77.7 Hz. As expected, the ^{19}F-NMR spectrum of $(CF_3)_2C=NNSF_2$ (17) consists of a complex multiplet of 16 lines for the NSF_2 group and of 2 x 12 lines for the CF_3 groups. This result can be interpreted as follows: One CF_3 is fixed in a trans position, the other in a cis position to the NSF_2 group. Coupling of $-NSF_2$ with one CF_3 group results in a quadruplet, further coupling with the second CF_3 group increases the multiplet splitting to 16 lines. Similar splitting of the two different CF_3 signals occurs: First, spin-spin coupling with the neighboring CF_3 group, and second, with the NSF_2 group. Another complex multiplet is shown by $(CF_3)_2CFNSF_2$ which has an unusual shift of 144.7 ppm for CF. Fifteen of the twenty-one possible peaks of this multiplet have been observed. ^{19}F-NMR spectra of F_5SNSF_2 are discussed in a later section. The ^{31}P-NMR spectrum of (24) has not been investigated. Fluorine atoms of the NSF_2 groups of compounds (8)- (13) are not equivalent with the exception of compounds of the composition RCF_2NSF_2 where coupling constants lie between 3-20 Hz.

TABLE 3

Data of Compounds Containing NSF$_2$ Groups

No.	Compound	δ_{S-F} (ppm)	ν_{S-F} (cm^{-1})	$\nu_{N=S}$ (cm^{-1})	References
1	Hg(NSF$_2$)$_2$	not det. (−)	680(vs)ν_{as}, 574(s)ν_s	1313(vs)	[15, 16]
2	F−NSF$_2$	−16.7 (2)	712(vs)ν_{as}, 615(m)ν_s	1150(m)	[6]
3	Cl−NSF$_2$	not det. (1)	792(vs)ν_{as}, 692(vs)ν_s	1200(s)	[16]
3a	Cl−NSF$_2$	−46.3 (1)	744(vs), 687(vs)	1189(s)	[8]
4	Br−NSF$_2$	−57.6 (1)	745(vs)ν_{as}, 687(vs)ν_s	1215(s)	[16]
4a	Br−NSF$_2$	not det. (?)	745(vs) ?, 685(s) ?	1215 ?	[8]
5	J−NSF$_2$ (?)	not det. (?)	735(vs)ν_{as}, 675(vs)ν_s	1235(s)	[16]
6	N≡C−NSF$_2$	−48.1 (1)	764(vs)ν_{as}, 725(vs)ν_s	1337(vs)	[21]
6a	N≡C−NSF$_2$	−48.7 (1)	765(vs)ν_{as}, 725(vs)ν_s	1348(vs)	[66]
7	CH$_3$−NSF$_2$	−69.9 (1)	858(s), 708(vs)	1357(vs)	[23, 50]
8	CH$_2$Cl−CH$_2$−NSF$_2$	−65.7 (cplx.)	639ν_{as}, 710.5ν_s	1383	[51]
9	CHCl$_2$−CH$_2$−NSF$_2$	−64.8 (cplx.)	646ν_{as}, 720ν_s	1380	[51]
10	CCl$_2$Br−CH$_2$−NSF$_2$	−65.6 (cplx.)	622ν_{as}, 695ν_s	1376.5/1340	[51]
11	CHClBr−CHCl−NSF$_2$	−58.0 (cplx.)	645ν_{as}, 719ν_s	1333	[51]
12	CCl$_2$Br−CHCl−NSF$_2$	−56.9 (cplx.)	650ν_{as}, 712ν_s	1343	[51]
13	CCl$_2$Br−CCl$_2$−NSF$_2$	−48.9 (?)	not det.	1368	[51]
14	CF$_3$−NSF$_2$	−52.1 (?)	714(m), 760(m)	1387(vs) ?	[13, 19, 48]
14a	CF$_3$−NSF$_2$	not det.	715(m)ν_{as}, 817(m)ν_s	1388(vs)	[52]
15	C$_2$F$_5$−NSF$_2$	−55.3 (?)	758(s) ?, 714(s) ?	1342(s) ?	[18, 19, 48]

TABLE 3 (Continued)

Data of Compounds Containing NSF$_2$ Groups

No.	Compound	δS-F (ppm)	νS-F (cm^{-1})	νN=S (cm^{-1})	References
16	CF$_2$(NSF$_2$)$_2$	-49.9 (3)	751(vs) ?, 705(vs) ?	1353(vs) ?	[22]
16a	CF$_2$(NSF$_2$)$_2$	-49.5 (3)	751(vs)ν_{as}, 705(vs)ν_s	1365(vs)	[66]
17	(CF$_3$)$_2$C=N-NSF$_2$	-31.6 (16)	759(s)ν_{as}, 706(s)ν_s	1364(s)	[46]
18	CF$_3$S-NSF$_2$	-59.8 (1)	736(vs) ?, 665(vs) ?	1287(s) ?	[39]
19	(CF$_3$)$_2$C(F)NSF$_2$	-58.2(14)	758(vs)ν_{as}, 708(vs)ν_s	1415(vs)	[46]
20	(CF$_3$)$_2$C(NSF$_2$)$_2$	-62.8 (7)	757(s)ν_{as}, 694(s)ν_s	1388(s)	[46]
21	[(CF$_3$)$_2$(F$_2$SN)C-N=]$_2$S	-66.2 (7)	732(ss)ν_{as}, 600(ss)ν_s	1405(vs, broad)	[46]
22	O(F$_2$)SN-CF$_2$-NSF$_2$	-48.8 (3)	760(s), 702(s)	1370(vs)	[54]
23	FC(O)-NSF$_2$	-41.6 (1)	764(vs), 727(vs)	1350(vs)	[32]
23a	FC(O)-NSF$_2$	-42.5 (1)	752(s)ν_{as}, 721(vs)ν_s	1350(vs)	[55]
24	F$_2$(O)P-NSF$_2$	-57.1 (6)	798(s)ν_{as}, 731(s)ν_s	1352(vs)	[31]
25	F$_5$S-NSF$_2$	-53.7 (5)	760 ?, 715 ?	1320(s) ?	[9]
25a	F$_5$S-NSF$_2$	-54.8 (?)	760(s), 714(s)	1313(s)	[11]
26	C$_6$F$_5$-NSF$_2$	-66.3 (3)	717(m), 655(m)	1370(s)	[54]
27	CF$_3$C(O)-NSF$_2$	-41.6 (4)	803(s), 749(s)	1362(vs)	[53, 87]
28	(CF$_3$)$_2$C(NH$_2$)-NSF$_2$	-63.4 (7)	736(s), 671(s)	1325(s)	[53]
29	FS(O$_2$)-NSF$_2$	-40.0 (2)	795(s) ?, 735(s) ?	1258(s) ?	[27, 37]

TABLE 3 (Continued)

Data of Compounds Containing NSF_2 Groups

29a	$FS(O_2)-NSF_2$	-37.4 (2)	800(vs) ν_{as}, 736(vs) ν_s	1275(vs)	[55]
30	$ClS(O_2)-NSF_2$	-37.7 (1)	805(vs) ν_{as}, 728(vs) ν_s	1220(vs)	[55]
30x	$CF_3SO_2-NSF_2$	-42.8 (4)	808(s) ν_{as}, 745(s) ν_s	not to det.	[69]
30y	$O_2S(NSF_2)_2$	-41.5 (1)	790(s) ?, 710(s) ?	1215(vs) ?	[71]

a (No. 1-30y; numbers and/or question marks in parentheses: multiplicity; other question marks: doubtful assignments).

TABLE 4

Chemical Shifts of Groups Connected with the NSF$_2$ Group and Coupling Constants[a]

No.	Compound	t (°C)	δ(X) (ppm)	J_{X-F} (Hz)	δ(CF$_3$) (ppm)	δ(CF$_2$) (ppm)	J_{FF}(C-F) (Hz)	J_{FF}(CF-SF) (Hz)
2	F-NSF$_2$	-30	77.7(3)	44.6±0.3 (NF-SF)	--	--	--	--
7	CH$_3$-NSF$_2$	--	2.30(3)	9.5(H-F)	--	--	--	--
14	CF$_3$-NSF$_2$	--	--	--	48.1(?)	--	--	?
15	C$_2$F$_5$-NSF$_2$	--	--	--	87.6(?)	163.4(?)	?	?
16	CF$_2$(NSF$_2$)$_2$	-20	--	--	--	40.5(5)	--	14.6
16a	CF$_2$(NSF$_2$)$_2$	-50	--	--	--	40.6(5)	--	14.5
17	(CF$_3$)$_2$C=N-NSF$_2$	30	--	--	70.5(12) (trans) 60.8(12) (cis)	--	6.6(cis-trans)	1.7(trans) 3.8(cis)
18	CF$_3$S-NSF$_2$	--	--	--	51.9 (1)	--	--	--
19	(CF$_3$)$_2$CFNSF$_2$	30	144.7(15) (theor.21)	--	82.7 (6)	--	4.6	18.5(CF-SF) 3.4(CF$_3$-SF)
20	(CF$_3$)$_2$C(NSF$_2$)$_2$	30	--	--	78.9 (5)	--	--	5.7
21	[(CF$_3$)$_2$(F$_2$SN)C-N=]$_2$S	30	--	--	77.5 (3)	--	--	6.6

TABLE 4 (Continued)

Chemical Shifts of Groups Connected with the NSF$_2$ Group and Coupling Constants[a]

No.	Compound	t (°C)	$\delta(X)$ (ppm)	J_{X-F} (Hz)	$\delta(CF_3)$ (ppm)	$\delta(CF_2)$ (ppm)	$J_{FF}(C-F)$ (Hz)	$J_{FF}(CF-SF)$ (Hz)
22	O(F$_2$)SN-CF$_2$-NSF$_2$	-15	-46.7 (9)	9.3(CF$_2$/SF$_2$O) 1.0(SF$_2$O/NSF$_2$)	--	40.9(4)	--	14.7
23	FC(O)-NSF$_2$	--	-20.1 (1)	--	--	--	--	--
23a	FC(O)-NSF$_2$	-80	-19.5 (3)(CF) -35.7 (2)(SF$_2$)	--	--	--	--	4.0
24	F$_2$(O)P-NSF$_2$	30	69.9 (6)(F$_2$)	1026(F$_2$PO) 28(F$_2$PO/NSF$_2$)	--	--	--	--
25	F$_5$S-NSF$_2$	--	(δ_A -71.3 (9)) (δ -84.1(36))	J_{AB} 154.1 (NSF$_2$/B) 13.6	--	--	--	--
25a	F$_5$S-NSF$_2$		(δ^B_A -87.5 (?)) (δ^A_B -84.1 (?))	? ?	--	--	--	--
26	C$_6$F$_5$-NSF$_2$	30	147.8(ortho)(?) 164.5(meta) (?) 159.5(para)(9)	20.5(para/meta) 2.0(para/ortho) 13.0(ortho/NSF$_2$)	--	--	--	--
27	CF$_3$C(O)-NSF$_2$	30	--	--	79.6(3)	--	--	2.1
28	(CF$_3$)$_2$C(NH$_2$)-NSF$_2$	30	-2.7(1)	not obs.	81.2(3)	--	--	3.0
29	FS(O$_2$)-NSF$_2$	-50	-44.0(3)	9.0	--	--	--	--

29a	$FS(O_2)-NSF_2$	−30	−61.5(3)	9.0	−	−	−	−
30x	$CF_3SO_2-NSF_2$	−	−	−	80.4(3)	−	−	3.8

a No. 2–30x; numbers/or symbols and/or question marks in parentheses: multiplicity and/or coupling groups; other question marks = data not given.

B. Characteristic Frequencies

Preparation of most of the substances containing a NSF_2 group succeeds via

$$R - NH_2 + SF_4 \rightarrow RN{=}SF_2 + 2 \, HF$$

Therefore it is reasonable to say that the $RNSF_2$ type of molecules are derivatives of SF_4, which has tetrahedral structures of symmetry C_{2v} and the structure can be assumed to be of trigonal bipyramidal origin. Two fluorine atoms and the lone electron pair are in equatorial positions, while the remaining two fluorine atoms are in axial positions so that sp^3d- hybridization of sulfur takes place. In general, bond angles for this type of molecule vary between 90 and 120 deg, but for SF_4 one angle has been measured to be nearly 180 deg [72]. Symmetric and antisymmetric stretching vibrations have been located at 894 and 715 (ν_s) and 863 and 728 cm^{-1} (ν_{as}). Deformation frequencies are found at 557, 239, 534, and 463 cm^{-1}, while a twisting vibration appears at 401 cm^{-1} [72]. On the other hand, $O{=}SF_2$ absorbs at 1333 (ν_{NS}), 748 (ν_{as}(SF)), 530 (bending), and 390 (δ_{as}) cm^{-1} [72]. Compounds of the type $RNSF_2$ should absorb in the same region, which means that one band of medium to very strong intensity should be assigned to ν_s and another to ν_{as}. Bands of medium to very strong intensity have been found for all $RNSF_2$ molecules. $Hg(NSF_2)_2$ (1) has its symmetric stretching frequency at 574, the antisymmetric stretching frequency at 680 cm^{-1}, and other compounds absorb in the same region, too. For instance, Griffiths et al. [52] have given a complete assignment of the vibrations of CF_3NSF_2, (14). The following vibrations have been assigned to ν_{NS} 1388, ν_s 817 (SF), and ν_{as} 716 (SF) cm^{-1}. A discussion of cis-trans isomerism demonstrates that no rotational isomers are present. From the above examples it seems to be clear that two bands of from medium to very strong intensity in the interval 550-850 cm^{-1} are characteristic for S-F vibrations of the compounds in question and may be assigned to ν_s and ν_{as}, while ν_{NS} for these types of compounds appears between 1150-1410 cm^{-1}.

III. OTHER COMPOUNDS
CONTAINING TETRAVALENT SULFUR

A. NMR Data

In Table 5 there are collected data for several NSF compounds which do not contain an NSF_2 group. The ^{19}F-NMR spectrum of one of them, NSF (31) has been already discussed. Trimeric and tetrameric NSF have a normal chemical shift, each with a single signal centered at -29.3 and -37.3 ppm, respectively. A similar shift (-37.1 ppm) has been found for the cyclic compound (34) with three sulfur atoms of different oxidation states (6 and 4) in the ring. There are two substances, (35, 40x) with three fluorine atoms attached to sulfur. The single broad peak obtained for $(CH_3)_2NSF_3$ (35) was centered at -41.4 ppm. It should be possible to resolve this signal on cooling because proton-fluorine coupling is expected. The proton NMR spectrum of (35) consists of a triplet from which a coupling constant J_{HF}=9.5 Hz was obtained. From the ^{19}F-NMR spectrum it is deduced [68] that the $(CH_3)_2N$ group occupies an equatorial position in the trigonal bipyramidal molecule. Considering the moderately broad signal of this molecule (from 20° to nearly -100°C), one can suppose that a very low energy barrier to fluorine exchange exists between axial and equatorial positions of the fluorines. At -84°C, shifting of the broad signal of $(C_2H_5)_2NSF_3$ (40x) to two signals of intensity 2:1 is observed. The signals are centered at -54.1 and -37.2 ppm. From this result it is deduced that two fluorine atoms occupy axial positions, while the third fluorine, a lone electron pair, and the amino group occupy equatorial positions of a trigonal bipyramidal structure as has already been found for compound (35). It was not possible to resolve the signals as would have been expected [86] for an A_2B type of spectrum. The three-membered heterocyclic compound (40), was first described [63] as a four-membered ring, but later

TABLE 5

Data of Compounds Containing Tetravalent Sulfur (without NSF$_2$ Compounds)[a]

No.	Compound	δ_{S-F} (ppm)	ν_{S-F} (cm^{-1})	$\nu_{N=S}$ (cm^{-1})	References
31	NSF	−240.3(1)	640(vs)	1372(s)	[1-3,17]
32	F–S=N / N=S–N / SF (ring)	−29.3(1)	720(s)? 650(s)?	1085(s)?	[3,4,17,28]
33	FS–N=SF / N=N / FS=N–SF	−37.3(1)	760(s)? 709(s)?	1117(s)?	[3,17]
34	O=N–SF=N / Cl–S / N=SF	−37.1(1)	799(vs)	?	[25,43]
35	$(CH_3)_2NSF_3$	−41.4(1)	993(vs)	–	[24,68]
36	F–N=S=O			1007(vs)	[20,84]
37	FC(O)–N=S–F / CF$_3$	−8(2)	771(s)? 712(s)?	1250(vs)	[58]
38	CF$_3$–N=S–F / CF(CF$_3$)$_2$	10.0(?)	725(m)? 705(s)?	1300(vs)?	[63]
39	C$_2$F$_5$–N=S–F / CF(CF$_3$)$_2$	24.3(?)	760(m)? 700(m)?	1300(vs)?	[63]

No.	Compound					Ref.
40	$C_2F_5C{=}N{\diagdown}{\diagup}S{-}F$, $CF(CF_3)_2$	31.5(?)	755(m)? 716(s)?		—	[63,64]
40x	$(C_2H_5)_2NSF_3$	– 46.4(1)	797(s)	739(vs)	—	[86]
40y	$NC{-}NS(F)N(CH_3)_2$	– 34.4(7)	728(s)	1218(vs)		[86]
40z	$NC{-}NS(F)N(C_2H_5)_2$	– 49.2(5)	685(m)	1180(s)		[86]

a No. 31–40 z; numbers or question marks in parentheses: multiplicity; other question marks = assignments doubtful; 35: at 25°C; at –100°C two signals of intensity 2:1 and δ_F = –59.4, –30.4; no H–F coupling observed; 36: t = 20°, δ_F = –123.3 ppm (3), J_{FN} = 112 Hz, ν(S=O) = 1246(vs), ν(N–F) = 834(s), from (84) ν(NS) = 997(s), ν in cm^{-1}; 37: t = –60°C; (FC(O)/SF)J_{FF} = 19 Hz, δ(FC(O)) = 50 ppm (2), δ(CF$_3$) = 71 ppm (4), (FC(O)/CF$_3$)J_{FF} = 0.54 Hz, (SF/CF$_3$)J_{FF} = 1.48 Hz; 38; δ(CF$_3$N) = 51.5 ppm (?), δ(CF) = 168.6 ppm (?), δ(CF$_3$)$_2$C = 70.3 ppm (?), no coupling constants given; 39: δ(CF$_3$C) = 91.1 ppm (?), δ(CF$_2$N = 94.0 ppm (?), δ(CF) = 166.6 ppm (?), δ[(CF$_3$)$_2$C] = 68.2 ppm (?), no coupling constants given; 40: δ(CF$_3$C) = 84.3 ppm (?), δ(CCF$_2$C) = 119.3 ppm (?), δ(CF) = 163.6 ppm (?), δ[(CF$_3$)$_2$C] = 72.2 ppm (?), no coupling constants given; 40x: δ(CH$_3$) = –1.27 ppm (3), δ(CH$_2$) = –3.35 ppm (4), J_{HH} = 6.9 Hz; 4Oy: t = 30°C, δ(CH$_3$) = –3.17 ppm (2), J_{FH} = 6.5 Hz; 40z: t = 30°C, δ(CH$_2$) = –3.75 ppm (8), δ(CH$_3$) = –1.40 ppm (3), J_{HF} = 5.5 Hz, J_{HH} = 6.9 Hz.

on a three-membered structure seemed to be more probable [64] to the authors. For none of the compounds (<u>38</u>), (<u>39</u>), or (<u>40</u>) have multiplicity and coupling constants been reported, perhaps because of the complexity of the ^{19}F-NMR spectra.

B. Characteristic Frequencies

A nonlinear structure (symmetry C_s) has been proposed for NSF. There are two structural possibilities [72] for this three-atomic molecule

III   IV

From microwave spectra, valence force constants, and the fact that bond orders are lower for thio- than for oxy-compounds, IV can be excluded. Because of this, the very strong band at 640 cm^{-1} is ν_{SF}, while the absorption at 1372 cm^{-1} belongs to ν_{NS}. The very simple ir spectrum of (NSF)$_3$, (<u>32</u>), seems to indicate a planar ring structure of symmetry D_{3h} for this substance. The lower vibration at 650 cm^{-1} may be assigned tentatively to ν_{SF}, while the absorption at 1080 cm^{-1} is the degenerated vibration ν_6 (E') of the ring. Strong delocalized π bonds have been concluded from this high value of 1080 cm^{-1} [72]. The ir spectrum of (NSF)$_4$ may be interpreted similarly, although it should be noted that this molecule is nonplanar.

For the discussion of the ir spectrum of FNSO see the general section about characteristic frequencies. No assignments of bands for compounds (<u>38</u>), (<u>39</u>), and (<u>40</u>) have been given.

IV. COMPOUNDS CONTAINING NSOF$_2$ GROUPS, SULFANURIC FLUORIDES, AND DERIVATIVES THEREOF

A. NMR Data

The main feature of these compounds is the oxidation number of 6 for sulfur compared to compounds with sulfur in the oxidation state 4 discussed in the two previous sections. Linear $NSOF_2$ compounds have a chemical shift ranging from -37.3 to -58.7 ppm, while sulfanuric fluorides and its derivatives show signals in the interval -70.9 to -78.8 ppm. It is not easy to understand why no N–F coupling could be observed for $FNSOF_2$ (41). It may be possible that p_π–d_π interactions between nitrogen and sulfur, or other disruptions of the electronic environment of nitrogen or fluorine play an important role in this. The J_{F-F} in $FNSO_2F$ is 24 Hz, while the shift of substances of the type $HalNSOF_2$ (Hal=F, Cl, Br) becomes approximately 7 ppm less negative in going from bromine to fluorine.

Several mercury–nitrogen–sulfur–fluorine compounds have been synthesized. The first had the composition $Hg[N(CF)_3SF_5]_2$, bp 44°C/3 Torr, a shift of -82.9 ppm for the AB_4 type of spectrum (SF_5 group), and -20.4 ppm for the CF_3 group, but no detailed analysis of the NMR spectrum has been given [61]. Another interesting mercury salt prepared by Glemser and his co–workers is $Hg(NSF_2)_2$. A ^{19}F–NMR spectrum has not been published (for other data see Table 1), probably because of the sensitive character of $Hg(NSF_2)_2$. Recently Seppelt and Sundermeyer [92] have been able to synthesize $Hg(NSOF_2)_2$ in high yields. This compound has been characterized by its ^{19}F–NMR spectrum (in CH_2Cl_2), which consists of a single peak centered at -58.7 ppm, as well as by other data. This is the lowest downfield shift found for this type of compound, comparable only to the shift of $(CH_3)_3SiNSOF_2$ (-55.9 ppm). Shifts of the other groups connected to $NSOF_2$ vary in the usual range, e.g., $\delta(CF_2)$=40.9 and 46.5 ppm. Coupling is always observed between $NSOF_2$ groups and other parts of the molecule which contain fluorine: Sometimes F–F coupling takes place over four bonds (J_{FF} = 8 Hz, FO_2SNSOF_2).

Iminosulfur oxidifluoride and a series of organic derivatives $RNSOF_2$ [R=methyl–, phenyl–, HOC_6H_4–, $HOOC(CH_2)_5$–] have been prepared in high yields. Their ^{19}F–NMR spectra have been obtained; no data have been published, but it is said that the signals occur in the usual range. Infrared data are given [94, 95].

The structural difference between the two isomers of sulfanuric fluoride could be deduced from their ^{19}F-NMR spectra: A cis isomer, with fluorine on one side of the six-membered ring and the three oxygen atoms on the other, gives rise to only one signal at -70.9 ppm, while the trans isomer, where fluorine and an oxygen of the cis form have changed their positions, exhibits a complex AB_2 type of ^{19}F spectrum with nine lines of which eight could be observed. A coupling constant of 21.6 ppm between the two types of fluorine was obtained.

B. Characteristic Frequencies

Many of the molecules of the type $RNSOF_2$ were prepared according to the reaction scheme

$$RNH_2 + SOF_4 \longrightarrow RNSOF_2 + 2\,HF$$

Sulfuroxytetrafluoride has a trigonal bipyramidal structure (symmetry C_{2v}) with oxygen in an equatorial position. Sulfur-fluorine stretching vibrations have been assigned to the bands at 820 (vs) and 927 (vs) cm^{-1}, while a band at 1379 (s) cm^{-1} was assigned to S=O stretch. Because of the spectral similarity of S=NH and S=O it is not possible to decide between ν (SO) and ν (SN) in $HNSOF_2$ which were located at 1200 (s) and 1428 (s) cm^{-1} [94]. In the region of S-F stretching vibration, one band of strong intensity is found at 840 cm^{-1} [94]. In this interval S-F vibrations of every molecule have been found (see Table 6). If assignments were made, the higher frequency was always assigned to ν_{as} and the lower to ν_s. Sometimes only one band of strong to very strong intensity appears between 740 and 930 cm^{-1}, note, e.g., $CINSOF_2$. In $NCNSOF_2$, ν_{as} is shifted to 927 cm^{-1}, while ν_{as} for all the other $NSOF_2$ compounds the frequency is found below 880 cm^{-1}. Only one band of $F_2SN-CF_2-NSOF_2$ of very strong intensity has been assigned to ν (OS-F) found at 843 cm^{-1}, while ν (S-F) appears at 760 (s) and 702 (s) cm^{-1}. Similar assignments to ν_{NSO} vary in the range 1160-1500 cm^{-1}. In a few cases, decisions could be made between ν (SO) and ν (NS). Thus, for instance,

ν(NS) in $F_2SNCF_2NSOF_2$ appears at 1370 (vs) cm^{-1} and data of $Hg(NSOF_2)_2$ may be taken from Table 6. A nonplanar ring structure is indicated by both ^{19}F-NMR and ir spectroscopy for sulfanuric fluorides, which do exist as cis and trans isomers. As has been mentioned above, cis-sulfanuric fluoride has its fluorine atoms on one side of the ring and the three oxygen atoms on the other. Thus, a higher symmetry (C_{3v}, degeneration of normal modes) results for the cis form than for trans-sulfanuric fluoride, which has two fluorine atoms and one oxygen on one side of the ring. Therefore, trans-sulfanuric fluoride gives rise to an ir spectrum with twice as many bands as the cis form. Infrared spectra of the derivatives of the two sulfanuric fluorides were run [56] and the bands are published but no detailed assignments were given. For further discussion of the infrared spectra of sulfanuric compounds see Banister et al. [98, 99].

V. COMPOUNDS CONTAINING NSO_2F GROUPS

A. NMR Data

Numerous substances of this type have been prepared by Roesky et al. ^{19}F-NMR absorptions take place in the region between -39.7 to -62.2 ppm relative to $CFCl_3$ (see Table 7). Lustig et al. [35] have synthesized F_2NSO_2F (79), which has been found to give rise to a single peak centered at -24.6 ppm, while F_2SNSO_2F marks the other end of the scale (-62.2 ppm). But this value (-24.6 ppm) is an exception, whereas the normal resonance absorption at this end of the scale occurs around -40 ppm. Although N-F coupling in F_2NSO_2F is expected, no multiplet can be observed. Similarly, this compound does not show F-F coupling.

But such a coupling should be possible because F_2NSO_2F is comparable with $F_2S(NSO_2F)_2$ (82), which shows splitting of the SO_2F signal into a triplet with a spacing of 6.9 Hz. Lustig et al. [35] also prepared $FN(SO_2F)_2$ (80).

Again this compound shows no coupling between N-F and F-F. Its chemical
shift of -44.9 ppm demonstrates that the fluorine of the SO_2F group is less
shielded than in F_2NSO_2F. Even stronger are the chemical shifts of the N-F
groups: -41.3 ppm (F_2N) and + 28.5 ppm (FN). The large differences in the
field positions of the S-F and N-F resonance bands were not explained by the
authors [35]. Compounds of similar composition are $HFNSO_2F$ (96) and
F_5SFNSO_2F (97). No H-F coupling between H and F at the nitrogen atom nor
between the nitrogen directly bonded to fluorine could be observed. Coupling
between the two fluorine atoms of $HFNSO_2F$ takes place with a coupling constant of 7.7 Hz. Details such as coupling constants or multiplicity about the
^{19}F-NMR spectrum of F_5SFNSO_2F were not published. Trichlorophosphazo-
sulfurylfluoride hydrolyzes to H_2NSO_2F (70, 70a), which is found to exhibit a
triplet on high resolution in the ^{19}F-NMR spectrum centered at -57.0 ppm. A
coupling constant of 6 Hz for proton-fluorine has been determined (see Table 8).

An interesting series of sulfuryl fluorides are the compounds $F_3P=NSO_2F$
(77), $Cl_3P=NSO_2F$ (76), and $Br_3P=NSO_2F$ (105). ^{19}F signals of the sulfuryl
group are centered at -60.9, -59.6 and -61.4 ppm, respectively. One would
expect a linear decrease of δ_F in this series instead of a more positive shift
for the trichlorophosphazo compound. In contrast to the chemical shifts,
coupling constants for P-F are lowered regularly for the same series: 16,
4, and 2 Hz. From electronegativity considerations, $F_3P=NSO_2F$ would be
expected to have the strongest downfield shift, but p_π-d_π interactions between
phosphorus and chlorine or fluorine may compensate the electronegativity
effect in the way that the lone electron pair at nitrogen is less affected than
in $Br_3P=NSO_2F$, and a positive shift compared to $Br_3P=NSO_2F$ for the
chlorine compound results. In $F_3P=NSO_2F$ the combined electronegativity
of three fluorine atoms is so much greater than that of three chlorine atoms
that the fluorine bonded to S is less shielded than in $Cl_3P=NSO_2F$. It may
be noted that organo-substituted NSO_2F molecules in general do not couple
the fluorine of the NSO_2F group with the organic part of the molecule. But
in case of the isomeric compounds (98-101), the N-methyl substituted
isomer always shows H-F coupling between this methyl group and fluorine,

J_{H-F}=1.4 Hz for both (see Table 8). An interesting feature of these compounds
is the positive value of the shift (in comparison to the other isomer) of the N-
methyl substituted molecules, which may be due to a positive inductive effect
of a methyl group in the vicinity of the NSO_2F group.

B. Characteristic Frequencies

The compounds of Tables 6 and 7 may be assumed to be derivatives of
sulfuryl fluoride; therefore it is reasonable to discuss the infrared spectrum
[72] of SO_2F_2. Sulfuryl fluoride belongs to the class of tetrahedral molecules
of symmetry C_{2v}. Bands of the S-O vibration appear at 1502 (ν_{as}) and
1269 (ν_s) cm^{-1} (=1385 cm^{-1}). Fluorine-sulfur vibrations have been found at
728 (ν_{as}) and 715 (ν_s) cm^{-1}. The high value of the S-O vibration depends on
p_π-d_π bonding [72] between sulfur and strong electronegative ligands. It is
known that sulfur in the case of tetrahedral coordination is able to use two
of five of its d orbitals for bonding. Furthermore, the S-O vibration (arith-
metical mean of ν_{as} + ν_s) in mixed compounds SO_2F-R has its absorption
bands in the range between the S-O absorptions of SO_2F_2 and SO_2R_2 [72].
Compare for example the series $SO_2(NSF_2)_2$, (30y), $SO_2(F)NSF_2$ [58], and
SO_2F_2.
   In Table 7 only a few assignments have been given, but in every case
ν_{as} is below the ν_{as} of SO_2F_2. Other values, which were not assigned in
the original paper, were assigned to these vibrations because of their
intensity and position. These values also fit in this series. Sulfur-fluorine
(80) vibrations are shifted to higher wavenumbers. Compound $FN(SO_2F)_2$,
(80), has very strong S-F absorptions, nearly 200 wavenumbers higher than
SO_2F_2. A strong electronegative group may be responsible for this shift.
A similar behavior can be observed by $F_2S(NSO_2F)_2$, (82), or F_2NSO_2F, (79).
The tentatively assigned S-F vibration of the isomeric compounds [98-101]
show a low and a higher value. If the assignments are correct, a relation
between ν_{S-F} and δ_{S-F} seems to exist for these substances: Compound (98)

TABLE 6

Data of Compounds Containing NSOF$_2$ Groups and of Sulfanuric Fluoride and Its Derivatives[a]

No.	Compound	δ_{S-F} (ppm)	ν_{S-F} (cm^{-1})		ν_{NSO} (cm^{-1})		References
41	FNSOF$_2$	−37.3 (2)	843(vs) ?	777(s) ?	1400(vs) ?	1120(vs) ?	[8]
42	ClNSOF$_2$	−43.0 (1)	821(vs) ?		1409(vs) ?	1158(vs) ?	[8]
42x	BrNSOF$_2$	−50.8 (1)	not det.		not det.		[96]
42y	Hg(NSOF$_2$)$_2$	−58.7 (1)	−		1231(s) ν_{as}	1198(s) ν_s (NS)	[92]
					1397(vs) ν_{as}	1362(s) ν_s (SO)	
42z	CO(NSOF$_2$)$_2$	−43.7 (1)	866(vs) ν_{as}	807(s) ν_s	1426(s) ν_{as}	1270(vs) ν_s	[93]
43	COF−NSOF$_2$	−46.0 (2)	811(m) ?	787(m) ?	1433(vs) ?	1164(m) ?	[8]
43z	NC−NSOF$_2$	−47.8 (1)	927(s) ν_{as}	868(vs) ν_s	1461(s) ν_{as}	1297(s) ν_s	[97]
44	CF$_2$(NSOF$_2$)$_2$	−46.4 (3)	858(vs)	811(s)	1447(vs) ν_{as}	1323(vs) ν_s	[30]
44z	F$_2$SN−CF$_2$−NSOF$_2$	−46.7 (9)	843(vs)		1465(vs) ν_{as}	1315(vs) ν_s	[54]
45	C$_6$F$_5$NSOF$_2$	−45.7 (6)	871(s)	808(vs)	1429(vs)	1268(vs)	[30]
45x	Cl$_2$SNCF$_2$NSOF$_2$	−49.6 (3)	826(vs) ?	742(s) ?	1456(vs) ν_{as}	1280(vs) ν_s	[88]

No.	Compound						Ref.
45y	$FO_2S-NSOF_2$	-45.4 (2)	888(s)?	821(s)?	1475(s)?	1245(s)?	[31]
45z	$(CH_3)_3SiNSOF_2$	-55.9 (1)	852(vs) ν_{as}	819(vs) ν_s	1495(vs)(SO)	1275(s)(NS)	[92]
46	cis-$(NSOF)_3$	-70.9 (1)	875(vs)?	776(vs)?	1395(vs)?	1168(vs)?	[26,27]
47	trans-$(NSOF)_3$	-71.86 (F,cplx.) -71.42 (F_2,cplx.)	899(vs)?	798(vs)?	1389(vs)?	1172(vs)?	[26,27]
48	$N_3S_3O_3F_2(C_6H_5)$	-77.2 (1)	810(s)		1155(vs)?	1140(s)?(NS)	[56,57]
49	$N_3S_3O_3F(C_6H_5)_2$	-72.0 (1)	775(m)		?	1135(vs)?(NS)	[56,57]
50	$N_3S_3O_3F(mo)_2$	-73.0 (1)	850(m)?	790(m)?	1160(s)?	1125(vs)?(NS)	[56,57]
51	$N_3S_3O_3F(mo)_2$	-73.2 (1)	840(s)?	780(m)?	1170(vs)?	1140(s)?(NS)	[56,57]
52	$N_3S_3O_3F(dimo)_2$	-72.6 (1)	850(m)?	790(m)?	1170(s)?	1130(vs)?(NS)	[56,57]
53	$N_3S_3O_3F(dimo)_2$	-78.1 (1)	850(m)?	790(m)?	?	1130(vs)?(NS)	[56,57]

TABLE 6 (Continued)

Data of Compounds Containing $NSOF_2$ Groups and of Sulfanuric Fluoride and Its Derivatives[a]

No.	Compound	δ S-F (ppm)	ν S-F (cm^{-1})			ν NSO (cm^{-1})	References
54	$N_3S_3O_3F(pip)_2$	-72.4 (1)	725(s) ?	835(s) ?	?	1135(vs) ? (NS)	[56, 57]
55	$N_3S_3O_3F(pip)_2$	-78.8 (1)	735(s) ?	840(s) ?	?	1125(vs) ? (NS)	[56, 57]
56	$N_3S_3O_3F(pyrro)_2$	-72.3 (1)	740(s) ?	845(s) ?	1170(vs) ?	1130(s) ? (NS)	[56, 57]
57	$N_3S_3O_3F(pyrro)_2$	-76.9 (1)	740(s) ?	845(s) ?	1145(s) ?	1130(s) ? (NS)	[56, 57]

[a] No. 41–57; numbers and/or symbols and/or question marks in parentheses: multiplicity or coupling group; other question marks = doubtful assignments; 41: δ (NF) = 113.1 ppm (3), J_{FF} = 24 Hz, 42y: in CH_2Cl_2; 42z: t = 30°C; 43: δ (CF) = -15.8 ppm (broad), at -95°C splitting into a triplet, J_{FF} = 10 Hz; 44: t = 30°C, $\delta(CF_2)$ = 40.9 ppm (5), J_{FF} = 9.5 Hz; 44z: δ (NSF_2) = -48.8 ppm (3, broad), $\delta(CF_2)$ = 40.9 ppm (9), $J_{FF}(NSF_2/CF_2)$ = 14.7 Hz, $J_{FF}(CF_2/SOF_2)$ = 9.3 Hz, $J_{FF}(NSF_2/SOF_2)$ = 1.0 Hz; 45: t = 30°C, δ (F, para) = 161.8 ppm (9), δ (F, meta) = 165.5 ppm (complex), δ (F, ortho) = 151.7 ppm (complex), $J_{FF}(SF_2/ortho)$ = 8 Hz, J_{FF}(para/meta) = 18.4 Hz, J_{FF}(para/ortho) = 1 Hz, at t = 120°C δ (SF_2) = -46.1 ppm (3), below 115°C F-atoms of SF_2 are not equivalent with J_{FF} = 0.9 Hz; 45x: $\delta(CF_2)$ = 46.5 ppm (3), J_{FF} = 9 Hz; 45y: δ (SF) = -60.1 ppm (3), J_{FF} = 8 Hz; 45z: δ_H not

det.; 46: $\delta_F = -72.25$ ppm in CH_3CN; 47: trans-$(NSOF)_3$: AB_2-spectrum, $J_{FF} = 21.6$ Hz, δ (F) $= -71.86$ ppm and δ (F_2) $= -72.76$ ppm in CH_3CN, $J_{FF} = 21.6$ Hz; 48-57: mo = morpholino, dimo = 2.6-dimethylmorpholino, pip = piperidino and pyrro = pyrrolidino derivatives of trimeric sulfanuric fluoride, IR data given without assignment for each compound.

TABLE 7

Data of Compounds Containing NSO_2F Groups[a]

No.	Compound	δ S-F (ppm)	ν S-F (cm^{-1})	ν S-O (cm^{-1})	References
58	F_2SN-SO_2F	-62.2(3)	848(s) ?	1443(s) ν_{as} 1225(vs) ν_{as}	[27, 37]
58a	F_2SN-SO_2F	-44.0(3)	–		[27, 37]
59	$OC(HN-SO_2F)_2$	-50.2(1)	810(s) ?	1412(s) ? 1209(vs) ?	[37]
60	$CH_3O_2CNH-SO_2F$	-52.1(1)	798(ss) ?	1415(s) ? 1210(vs) ?	[37]
61	$C_2H_5O_2CNH-SO_2F$	-52.4(1)	795(s) ?	1400(s) ν_{as} 1205(vs) ν_s	[37]
62	$(C_2H_5)_2NCONH-SO_2F$	-49.6(1)	?	1450 ν_{as} 1210 ν_s	[37]
63	$(CH_3)_2S=N-SO_2F$	-59.0(1)	850(m) ?	1360 ν_{as} 1200 ν_s	[37]
64	$OCN-SO_2F$	-61.0(1)	?	1385(s) ν_{as} 1218(s) ν_s	[37]

No.	Compound	δ	IR	IR		Ref.
64a	$OCN-SO_2F$	−61.1(1)	833(s)?	1380(s)?	1235(s)?	[65]
65	$F(O)C(F_2SO)NSO_2F$	−54.6(4)	827(s)?	1495(s)?	1205(s)?	[65]
66	$F(O)C(F_3CO)N-SO_2F$	−50.4(8)	823(vs)?	1308(s)?	1206(vs)?	[65]
67	$(F(O)C-N-SO_2F)_2$	−57.4(1)	829(s)?	1284(s)?	1225(s)?	[65]
68	$OSN-SO_2F$	−59.2(1)	830(s)	$1435\,\nu_{as}$	$1260\,\nu_s$	[38]
69	Cl_2SN-SO_2F	−61.4(1)	805(s)	$1425\,\nu_{as}$	$1210\,\nu_s$	[38]
70	H_2N-SO_2F	−57.0(3)	?	?		[38]
70a	H_2N-SO_2	−57.3(1)	784(vs)?	1410(vs)?		[33, 79]
71	$Cl_2C=NSO_2F$	−54.2(1)	850(s)	$1440\,\nu_{as}$	$1220\,\nu_s$	[75]
72	$\overset{\displaystyle C(CH_3)C(CH_3)=CH_2}{\underset{\displaystyle CS(O)N(H)-SO_2F}{\|}}$	−49.6(1)	780(vs)	1435(vs)	1220(vs)	[76]

TABLE 7 (Continued)

Data of Compounds Containing NSO_2F Groups

No.	Compound	δ S-F (ppm)	ν S-F (cm^{-1})	ν S-O (cm^{-1})	References
73	$(CH_3)_2SiN(H)SO_2F$	-61.6(2)	850(vs) ?	1410(vs) ? 1210(vs) ?	[76]
74	F_2OSNSO_2F	-60.1(3)	821(s) ?	1450(s) ? 1245(s) ?	[31]
75	$CF_3-NH-SO_2F$	-55.4(8)	810(vs)	1460(vs) ν_{as} 1265(vs) ν_s	[36]
76	$Cl_3P=N-SO_2F$	-59.6(2)	-	-	[33]
77	$F_3P=N-SO_2F$	-60.9(8)	?	1434(vs) ν_{as} 1220(vs) ν_s	[34]
78	$HN(SO_2F)_2$	-57.8(1)	-	-	[33]
78a	$HN(SO_2F)_2$	-56.9(1)	?	?	[35, 80]
79	F_2NSO_2F	-24.6(1)	845(vs)	1483(vs) ν_{as} 1250(vs) ν_s	[35]
80	$FN(SO_2F)_2$	-44.9(1)	902(vs) ν_{as} 851(vs) ν_s	1500(vs) ν_{as} 1243(vs) ν_s	[35]
81	$S(NSO_2F)_2$	-58.5(1)	770-850(vs) ?	1450(vs) ? 1220(vs) ?	[44]
82	$F_2S(NSO_2F)_2$	-62.0(3)	780-900(vs) ?	1460(vs) ? 1220(vs) ?	[44]
83	$H_2NSOFNSO_2F$	-57.8(2)	848(vs) ? 795(vs) ?	1408(vs) ? 1225(vs) ?	[77]
84	$H_3C(H)NSOFNSO_2F$	-58.0(?)	840(vs) ? 785(vs) ?	1415(vs) ? 1210(vs) ?	[77]

No.	Compound	Shift					Ref.
85	(H₃C)HNSO₂NSO₂F	−58.6(2)	835(vs)?	787(vs)?	1410(vs)?	1210(vs)?	[77]
86	(H₅C₂)₂NSOFNSO₂F	−58.7(?)	834(vs)?	787(vs)?	1410(vs)?	1210(vs)?	[77]
87	(H₃C)₂NSF(NSO₂F)₂	−60.3(2)	770–870(vs)?		1410(s)?	?	[78]
88	[(H₃C)₂N]₂S(NSO₂F)₂	−58.7(1)	860(s)?		1392(vs)?	1206(vs)?	[78]
89	[(H₃C)₂N]₂SO(NSO₂F)₂	−58.5(1)	825(vs)?		1380(vs)?	1190(vs)?	[78]
90	H₂FCClCNSO₂F	−54.5(1)	825(vs)?		1430(vs)?	1218(vs)?	[60]
91	F₃COCNHSO₂F	−53.8(1)	813(s)?		1297(vs)?	1180(vs)?	[60]
92	F₃CClCNSO₂F	−55.2(1)	862(vs)		1468(vs) ν_{as}	1229(vs)?	[60]
93	H₂NF₃CNSO₂F	−51.6(1)	811(vs)?		1382(s)?	1198(vs)?	[60]
94	(C₂H₅)₂NF₃GCNSO₂F	−57.5(4)	850(vs)?		1431(m)?	1219(vs)?	[60]
95	(CH₃)₂NF₃CCNSO₂F	−57.0(4)	879(vs)?		1426(m)?	1197(vs)?	[60]
96	HFN-SO₂F	−39.6(2)	820(vs)?		1472(vs)?	1240(vs)?	[81]
97	F₅SFN-SO₂F	−41.0(?)	?		1490(vs)?	1250(vs)?	[81]
98	CH₃O—C=NSO₂F / (CH₃)₂N	−60.3(1)	735(vs)?		1495(vs)?	1240(s)?	[70]
99	(CH₃)₂N-C-NSO₂F ‖O CH₃	−42.4(4)	775(vs)?		1420(vs)?	1215(vs)?	[70]

TABLE 7 (Continued)

Data of Compounds Containing NSO_2F Groups

No.	Compound	δS-F (ppm)	νS-F (cm^{-1})	νS-O (cm^{-1})		References
100	CH_3O\C=NSO$_2$F	-59.5(1)	730(vs) ?	1400(vs) ?	$_o$1238(vs) ?	[70]
101	$(C_2H_5)_2N$\C-NSO$_2$F $\parallel$ O CH$_3$	-41.3(4)	759(vs) ?	1425(vs) ?	1215(vs) ?	[70]
102	$CH_3CH=NSO_2F$	-39.7(8)	772(vs) ?	1410(vs) ?	1210(vs) ?	[70]
103	$C_2H_5CH=NSO_2F$	-39.8(6)	792-797(vs) ?	1415(vs) ?	1210(vs) ?	[70]
104	$C_6H_5CH=NSO_2F$	-43.0(1)	756-781(vs) ?	1395(vs) ?	1200(vs) ?	[70]
105	$Br_3P=NSO_2F$	-61.4(2)	810(vs)	1380(vs) ν_{as}	1190(vs) ν_s	[83]
106	$CH_3CONHSO_2F$	-51.4(?)	795(vs) ?	1490(vs) ?	1250(m) ?	[83]
107	$CH_2ClCONHSO_2F$	-52.4(?)	818(vs) ?	1485(vs) ?	1232(vs) ?	[82]
108	$CHCl_2CONHSO_2F$	-54.9(?)	812(vs) ?	1400(s) ?	1222(vs) ?	[82]
109	$CCl_3CONHSO_2F$	-53.3(?)	848(vs) ?	1485(vs) ?	1241(s) ?	[82]
110	$CH_3CCl=NSO_2F$	-53.8(?)	830(vs) ?	1430(vs) ?	1214(vs) ?	[82]
111	$CH_2ClCCl=NSO_2F$	-54.9(?)	820(s) ?	1422(vs) ?	1213(vs) ?	[82]
112	$CHCl_2CCl=NSO_2F$	-56.0(?)	827(s) ?	1439(vs) ?	1220(vs) ?	[82]
113	$CCl_3CCl=NSO_2F$	-57.3(?)	803(vs) ?	1451(vs) ?	1229(vs) ?	[82]

No.	Compound					
114	$CH_3C(OCH_3)=NSO_2F$	-51.0(?)	789(vs)?	1400(vs)?	1212(vs)?	[82]
115	$CH_3C(OC_2H_5)=NSO_2F$	-51.1(?)	803(s)?	1391(vs)?	1205(vs)?	[82]
116	$CH_3C(NHCH_3)=NSO_2F$	-51.7(?)	862(s)?	1408(m)?	1183(vs)?	[82]
117	$CH_3C(NHC_2H_5)=NSO_2F$	-51.6(?)	782(vs)?	1350(vs)?	1170(vs)?	[82]
118	$CH_3C[N(C_2H_5)_2]=NSO_2F$	-52.6(?)	734(vs)?	1358(vs)?	1170(vs)?	[82]
119	$CH_3C(NH_2)=NSO_2F$	-50.5(?)	793(vs)?	1368(vs)?	1190(vs)?	[82]
120	$CH_3OC(NH_2)=NSO_2F$	-53.0(?)	770(s)?	1390(vs)?	1178(vs)?	[82]
121	$H_2N-C-NSO_2F$ $\parallel$ O CH_3	-49.2(4)	790(vs)?	1430(vs)?	1320(s)?	[82]

a No. 58–121; numbers and/or question marks in parentheses: multiplicity of the signal; other question marks = no or doubtful assignment.

TABLE 8

Additional Data for Compounds No. 58–121[a]

No.	Compound	t (°C)	$\delta(CF)$ (ppm)	$\delta(XF)$ (ppm)
58	F_2SNSO_2F	30	--	-40.0(2)
58a	F_2SNSO_2F	-50	--	-35.7(2)
59	$OC(HNSO_2F)$	--	--	--
60	$CH_3O_2CNHSO_2F$	melt	--	--
61	$C_2H_5O_2CNHSO_2F$	60	--	--
63	$(CH_3)_2S=NSO_2F$	--	--	--
65	$FOC(FO_2SO)NSO_2F$	--	6.58(4)	-43.6(4)
66	$FOC(F_3CO)NSO_2F$	--	6.6(5,CF) 68.4(4,CF_3)	--
67	$(FOC-N-SO_2F)_2$	--	9.08(1)	--
70	H_2NSO_2F	--	--	--
72	$\underset{CS(O)NHSO_2F}{\overset{C(CH_3)C(CH_3)=CH_2}{\|}}$	--	--	--
73	$(CH_3)_3SiNHSO_2F$	--	--	--
74	F_2OSNSO_2F	--	--	-45.4(2)
75	CF_3NHSO_2F	--	58.0(4)	--
76	$Cl_3P=NSO_2F$	--	--	--
77	$F_3P=NSO_2F$	--	--	86.7(4)
78	$HN(SO_2F)_2$	--	--	--
78a	$HN(SO_2F)_2$	--	--	--

J_{F-F} (Hz)	$\delta(CH_3)$ (ppm)	$\delta(CH_x)$ (ppm)	$\delta(NH)$ (ppm)	J_{Z-F} (Hz)
9	--	--	--	--
9	--	--	--	--
--	--	--	-9.2(1)	--
--	-3.77(1)	--	-8.62(1)	--
--	-1.1 (3)	-4.2(4)	-8.9(1)	--
--	-2.98(1)	--	--	--
11.4(CF/OSO_2F) 8.5(CF/SO_2F) 4.4(OSO_2F/SO_2F)	--	--	--	--
4.3(CF/CF_3) 1.8(SF/CF_3) 5.4(SF/CF)	--	--	--	--
--	--	--	--	--
--	--	--	6.1(2)	6
--	-10.2(?)	-12.5(?) (CH_2) -11.8(?) (CH)	-12.1(?)	--
--	0.33(2)	--	5.7(1)	6
8	--	--	--	--
6	--	--	-7.45(1)	4
--	--	--	--	4
4	--	--	--	16
--	--	--	-10.3(1)	--
--	--	--	-9.0	--

TABLE 8 (Continued)

Additional Data for Compounds No. 58-121[a]

No.	Compound	t (°C)	δ(CF) (ppm)	δ(XF) (ppm)
79	F_2N-SO_2F	-50	--	-41.3(1)
80	$FN(SO_2F)$	25	--	28.5(1)
82	$F_2S(NSO_2F)_2$	--	--	-45.3(3)
83	$H_2NSOFNSO_2F$	--	--	-68.2(6)
84	$H_3CHNSOFNSO_2F$	--	--	-57.0(?)
85	$(H_3C)_2NSOFNSO_2F$	--	--	-50.4(14)
86	$(H_5C_2)_2NSOFNSO_2F$	--	--	-62.0(?)
87	$(H_3C)_2NSF(NSO_2F)_2$	--	--	-58.4(21)
88	$[(H_3C)_2N]_2S(NSO_2F)_2$	--	--	--
89	$[(H_3C)_2N]SO(NSO_2F)_2$	--	--	--
90	$H_2FCClCNSO_2F$	--	211.8(3)	--
91	$F_3COCNHSO_2F$	--	77.0(1)	--
92	$F_3CClCNSO_2F$	--	73.7(1)	--
93	$H_2NF_3CCNSO_2F$	--	76.6(1)	--
94	$(C_2H_5)_2NF_3CCNSO_2F$	--	62.6(2)	--
95	$(CH_3)_2NF_3CCNSO_2F$	--	63.6(14)	--
96	$HFN-SO_2F$	--	--	91.8(2)
97	$F_5SFN-SO_2F$	--	--	23.2(?, NF) -61.7(?, SF) -69.0(?, SF$_4$)

J_{F-F} (Hz)	$\delta(CH_3)$ (ppm)	$\delta(CH_x)$ (ppm)	$\delta(NH)$ (ppm)	J_{Z-F} (Hz)
--	--	--	--	--
--	--	--	--	--
6.9	--	--	--	--
8.5	--	--	-6.9(2)	4.8 (NH/FSO)
8.5	-3.18(?)	--	-6.43(?)	3.0 (CH/FSO)
8.2	-3.25(?)	--	--	3.4 (CH/FSO)
8.0	-1.39(?)	-3.67(?)	--	2.8 (CH₂/FSO)
7.4	--	--	--	not det.
--	-2.90(1)	--	--	--
--	-2.99(1)	--	--	--
--	--	-5.14(3)	--	45.7 (CF/H)
--	--	--	-10.2(1)	--
--	--	--	--	--
--	--	--	-8.34(1)	--
8.5	-1.24(3)	3.60(4)	--	--
7.7	-3.25(4)	--	--	1.5 (CH/CF)
7.7	--	--	--	--
not det.	--	--	--	--

TABLE 8 (Continued)

Additional Data for Compounds No. 58-121[a]

No.	Compound	t (°C)	$\delta(CF)$ (ppm)	$\delta(XF)$ (ppm)
98	$\begin{array}{c}CH_3O\\ \diagdown\\ C=NSO_2F\\ \diagup\\ (CH_3)_2N\end{array}$	--	--	--
99	$(CH_3)_2N-\underset{\underset{O}{\|\|}}{C}-\underset{\underset{CH_3}{\|}}{N}SO_2F$	--	--	--
100	$\begin{array}{c}CH_3O\\ \diagdown\\ C=NSO_2F\\ \diagup\\ (C_2H_5)_2N\end{array}$	--	--	--
101	$(C_2H_5)_2N-\underset{\underset{O}{\|\|}}{C}-\underset{\underset{CH_3}{\|}}{N}SO_2F$	--	--	--
102	$CH_3CH=NSO_2F$	--	--	--
103	$C_2H_5CH=NSO_2F$	--	--	--
104	$C_6H_5CH=NSO_2F$	--	--	--
105	$Br_3P=NSO_2F$	66	--	--
106	$CH_3CONHSO_2F$	--	--	--
107	$CH_2ClCONHSO_2F$	--	--	--
108	$CHCl_2CONHSO_2F$	--	--	--
109	$CCl_3CONHSO_2F$	--	--	--
110	$CH_3CCl=NSO_2F$	--	--	--
111	$CH_2ClCCl=NSO_2F$	--	--	--
112	$CHCl_2CCl=NSO_2F$	--	--	--

J_{F-F} (Hz)	$\delta(CH_3)$ (ppm)	$\delta(CH_X)$ (ppm)	$\delta(NH)$ (ppm)	J_{Z-F} (Hz)
--	-3.28(1)(N) -4.12(1)(O)	--	--	--
--	-3.20(2)(NCH$_3$) -3.06(1) [N(CH$_3$)$_2$]	--	--	1.4 (NCH$_3$)
--	-1.33(3)(CH$_2$) -4.17(1)(0)	-- -3.65(4)	--	--
--	-1.22(3) (CH$_2$) -3.16(2)(N)	-3.43(4)	--	1.4 (NCH$_3$)
--	-2.26(4)	-8.65(8)	--	1.8 (CH$_3$) 0.8 (CH)
--	-1.08(6)	-2.59(8) (CH$_2$) -8.52(6) (CH)	--	2.0 (CH$_2$) 0.7 (CH)
--	--	-8.47(1) (CH) -7.15 (compl.)	--	--
--	--	--	--	2
--	-2.12(?)	--	-10.19(?)	--
--	--	-4.30(?)	-10.6(?)	--
--	--	-6.46(?)	-10.13(?)	--
--	--	--	-11.42(?)	--
--	-2.73(?)	--	--	--
--	--	-4.77(?)	--	--
--	--	-6.59(?)	--	--

TABLE 8 (Continued)

Additional Data for Compounds No. 58-121[a]

No.	Compound	t (°C)	δ(CF) (ppm)	δ(XF) (ppm)
114	$CH_3C(OCH_3)=NSO_2F$	--	--	--
115	$CH_3C(OC_2H_5)=NSO_2F$	--	--	--
116	$CH_3C(NHCH_3)=NSO_2F$	--	--	--
117	$CH_3C(NHC_2H_5)=NSO_2F$	--	--	--
118	$CH_3C[N(C_2H_5)_2]=NSO_2F$	--	--	--
119	$CH_3C(NH_2)=NSO_2F$	--	--	--
120	$CH_3OC(NH_2)=NSO_2F$	--	--	--
121	$H_2N-C-NSO_2F$ $\underset{\text{O\ \ CH}_3}{\overset{\|\ \ \ \ |}{}}$	--	--	--

[a]Meaning of Z depends on compound and may be H, N, P or F;
t = temperature of NMR measurements; numbers or symbols in
parentheses: multiplicity or shift of the special groups or coupling
groups; No. 61, 101, 103, 115, 118: $J_{CH_2-CH_3}$ = 7 Hz; No. 86:
$J_{CH_2-CH_3}$ = 7.3 Hz; No. 94: $J_{CH_2-CH_3}$ = 7.5 Hz; No. 102:
J_{CH-CH_2} = 5 Hz; No. 103: J_{CH-CH_2} = 3.2 Hz.

J_{F-F} (Hz)	$\delta(CH_3)$ (ppm)	$\delta(CH_x)$ (ppm)	$\delta(NH)$ (ppm)	J_{Z-F} (Hz)
--	-3.39(?, CH$_3$) -2.47(?,0)	--	--	--
--	-2.45(?, CH$_3$) -1.63(?, CH$_2$CH$_3$)	-4.33(?)	--	--
--	-2.94(?, (CH$_3$) -2.42(?,N)	--	-7.8(?)	--
--	-2.34(?, (CH$_3$) -1.14(?, (CH$_2$CH$_3$)	-3.33(?)	-7.73(?)	--
--	-2.57(?, CH$_3$) -1.32(?, C$_2$H$_5$)	-3.66(?)	--	--
--	-2.20(?)	--	-7.47(?)	--
--	-3.99(?)	--	-6.14(?)	--
--	-3.47(?)	--	-6.85(3)	--

gives rise to a resonance signal at –60.3 ppm while ν_{S-F} appears at 735 cm^{-1}, compound (99) absorbs at –42.4 ppm and ν_{S-F}=775 cm^{-1}, compound (100) shows a single peak at –59.5 ppm and ν_{S-F} is found at 759 cm^{-1}.

VI. COMPOUNDS CONTAINING NSF$_5$-GROUPS (TABLE 9)

A. NMR Data

^{19}F-NMR spectra of SF$_5$ compounds are often difficult to resolve and to assign. Since the five fluorine atoms are nonequivalent, spectra have to be interpreted as AB$_4$ type spectra. One of the best investigated spectra is that of F$_2$S=NSF$_5$ (123, 123a), for which Cohen, Hooper, and Peacock [9] observed a nine-line spectrum of the single fluorine of the molecule centered at –71.3 ppm, a multiplet of 36 lines at –84.1 ppm belonging to the four basal fluorines of the SF$_5$ group, and a quintuplet at –53.7 ppm due to the SF$_2$ group. A coupling constant of 154.1 Hz shows strong coupling between the A and B$_4$ part of the molecule, while coupling between the B$_4$ part and the SF$_2$ group results in a coupling constant of 13.6 Hz. There are different shifts δ_A published for this molecule, as is also the case for F$_2$NSF$_5$ (124, 124a, 124b) (see Table 9). Data from a series of NSF$_5$ compounds (129–132) have been obtained from the original paper [61] by the following procedure: The SF$_5$ part is characterized by the difference $\delta_A - \delta_B$ and the center of δ_A and δ_B. Thus it was possible to take half of $\delta_A - \delta_B$ and to add or to substract this value to or from the chemical shift of the center of δ_A and δ_B. Having done this, the values obtained were assigned to δ_A and δ_B by comparison with other shifts of δ_A and δ_B published in Ref. [42]. Similar evaluations of δ_A and δ_B are given for F$_2$NSF$_5$, (124a). Assignments are of tentative character. Other NSF$_5$ compounds described in Ref. [61] were not included in this collection, because either no or only a few NMR data are given.

TABLE 9

Data of Compounds Containing NSF_5 Groups[a]

No.	Compound	δ_A (ppm)	δ_B (ppm)	J_{AB} (Hz)	$J_{A,B-X}$ (Hz)	ν_{S-F} (cm^{-1})	δ_X (ppm)	References
122	FO_2SNFSF_5	-61.7(?)	-69.0(?)	not det.	not det.	?	23.2(?, NF) -41.0 (?, SO_2F)	[81]
123	F_2SNSF_5	-71.3(9)	-84.1(36)	154.1	13.6(BX)	?	-53.7(5)	[9]
123a	F_2SNSF_5	-87.5(?)	-84.1(?)	?	?	910,879 592	-54.8(?)	[11,12]
124	F_2NSF_5	-49.6(?)	-37.1(?)	153.2	?	912,885, 699,606, 569	-65.9(?)	[41]
124a	F_2NSF_5	-33.1(9)	-38.7(12)	144.0	?	?	-66.9(3)	[10]
124b	F_2NSF_5	-50.4 (cplx.)	-38.4 (cplx.)	not det.	4.5±0.3 (A/NF_2) 19(B/NF_2)	911,882?	-68.2(3)	[62]
125	Cl_2NSF_5	-59.4?	-61.4?	?	–	913,862, 600	–	[13]
126	CF_3NFSF_5	-58.2 not resolved	–	–	–	?	48.2(?, NF) 70.3(?, CF)	[18]

TABLE 9 (Continued)

Data of Compounds Containing NSF_5 Groups

No.	Compound	δ_A (ppm)	δ_B (ppm)	J_{AB} (Hz)	$J_{A,B-X}$ (Hz)	ν_{S-F} (cm^{-1})	δ_X (ppm)	References
127	$C_2F_5NFSF_5$	−60.5 not resolved		−	−	?	49.4(?, NF), 81.7(2, CF$_3$), 110.1(?, CF$_2$)	[18]
128	$F_5SNSF_2NSF_5$	?	?	?	?	920, 863, 592	?	[12]
129	$Cl_2C=NSF_5$	−67.6?	−72.2?	157.0	−	?	−	[61]
130	$(ClC=NSF_5)_2$	−63.2?	−66.6?	154.0	−	?	−	[61]
131	$SCNSF_5$	−67.5?	−82.3?	156.0	−	?	−	[61]
132	$OCNSF_5$	−66.3?	−83.6?	149.0	−	?	−	[61]
133	H_2NSF_5	not det.	not det.	not det.	not det.	930, 885, 694	not det.	[40]

a No. 122–133; numbers or symbols or question marks in parentheses: multiplicity or shift of the special group; other question marks: no or doubtful assignment; No. 124 and 124b: J_{N-F} not det., No. 124a: J_{N-F} = 120 Hz.

B. Characteristic Frequencies

Molecules containing an SF_5 group may be assumed to be derivatives either of SF_6 or SF_5Cl. Five vibrations ν_1, ν_2, ν_3, ν_4, and ν_5 are found in the spectrum of SF_6, while ν_6 is Raman- and ir-forbidden and can be estimated only from overtones or combination bands. SF_6 is a regular octahedron of symmetry O_h. Sulfurpentafluoride-chloride belongs to the point group C_{4v}. A very strong parallel band in the ir spectrum at 854 cm^{-1} is assigned to ν_1, the axial S-F stretching frequency. A very strong perpendicular ir band at 706 cm^{-1} is assigned to the symmetry SF_4-square stretching frequency $\nu_2(a_1)$ while another intense parallel band found in the ir spectrum is assigned to symmetrical out of plane SF_4-square deformation (ν_3). The highest frequency at 908 cm^{-1} is assigned to ν_8 (e) (SF_4-square stretch). Other frequencies are of limited value for our purpose. For literature references see Ref. [72].

The infrared spectra of compounds in Table 9 are complex and in most cases ir spectra were not recorded or bands were not assigned. In general, bands of from strong to very strong intensity in the range 600-900 cm^{-1} have been assigned to S-F frequencies of the SF_5 group. The absorption at the highest wavenumbers (930 cm^{-1}) due to S-F vibrations has been found for pentafluorosulfanylamine (133).

VII. OTHER COMPOUNDS
CONTAINING HEXAVALENT SULFUR

A. NMR Data

One of the most interesting N-S-F compounds, NSF_3 (134), has been synthesized by Glemser and his co-workers [1]. This chemically and thermally stable gaseous compound is the only example where nitrogen-fluorine coupling with J_{NF}=27 Hz over two bonds (a triple and a single bond) is observed. Such a coupling does not take place in derivatives of NSF_3, e.g., $NSF_2N(C_2H_5)_2$, (<u>136</u>, <u>136a</u>) or $NSF_2OC_6H_5$ (<u>137</u>). The resonance signal in these cases consists of one signal and is shifted around 10 ppm downfield from the multiplet of NSF_3. Less shielding of the remaining fluorine occurs, although it is expected that electron donating groups should effect better shielding. There are only a few compounds with fluorine in the vicinity of Si-N bonds. $[(CH_3)_3SiN]_2SF_2$ (<u>138</u>) has a chemical shift of -77. 4 ppm and fits well into the series of hexavalent sulfur compounds. N-F coupling (J_{NF}=120 Hz) is found again in the case of fluorine directly bonded to nitrogen. The structure of the molecule in question, $CF_3SF_4NF_2$ (<u>142</u>) has been deduced from its ^{19}F-NMR spectrum to be a tetragonal bipyramide where CF_3 and NF_2 groups are situated at the two apexes. A very complex ^{19}F-NMR spectrum was obtained for $CF_3OSF_4NF_2$ (<u>143</u>). Coupling of N-F is of the usual order of magnitude: J_{NF}=110 Hz. Because of the great number of the coupling nuclei (see Table 10) a longer set of coupling constants were determined and it is not possible to include all these data in the table. As a result of shifts and observed coupling, the structure given in the table was deduced. Compounds (<u>142</u>) and (<u>143</u>) may be compared in that both are tetragonal bipyramides, but in (<u>142</u>) the CF_3 and NF_2 are located in a trans position, while in (143) CF_3O and CF_3 are in a cis position.

In general most of the values of the chemical shifts of these hexavalent sulfur compounds are in the range of -45 to -80 ppm. This means that the fluorine atoms in these compounds are less shielded (in general) as compared to the other N-S-F molecules.

B. Characteristic Frequencies

Thiazyl trifluoride NSF_3 (<u>134</u>, <u>134a</u>) was compared [2, 3, 72] with OPF_3. Both are tetrahedral molecules of symmetry C_{3v} and both have similar valence force constants (f_{OP}=11. 38, f_{NS}=12. 55, f_{PF}=6. 35, f_{SF}=4. 49) and similar stretching frequencies, e.g., ν_1 (a$_1$)=1415 (OP), 1515 (NS) cm^{-1} or ν_2(A$_1$)=873 (PF), 775 (SF) cm^{-1} or ν_4(e)=990 (PF), 811 (SF) cm^{-1}. From f$_{NS}$ the bond order [72] has been estimated to be 2.8, which means that the N-S bond has the character of a triple bond. Similar absorption frequencies appeared in the infrared spectrum of NSF$_2$N(C$_2$H$_5$)$_2$ (136). In this case the S-F vibration was found at 811 (ν_{as}) and 755 (ν_s) cm^{-1}. Vibrations at 883 and 847 cm^{-1} in the ir spectrum of (CF$_3$N)$_2$SF$_2$, have been assigned to ν_{SF}. The perfluoroethyl compound (140) does absorb in the same range. A very broad band of very strong intensity in the interval 780-900 cm^{-1} could not be resolved between ν_{as} and ν_s.

VIII. MISCELLANEOUS COMPOUNDS

A. NMR Data

Nothing can be said about the range where the ^{19}F-NMR absorption of compounds in Table 11 does occur. But some interesting details were observed running the spectra. CF$_3$SNF$_2$ (144) synthesized by Stump and Padgett [59] is the only compound with bivalent sulfur between a trifluoromethyl and a difluoro amino group. N-F coupling does not occur and only a broad singulet is observed for the NF$_2$ group. But the ^{19}F-NMR signal of the CF$_3$ group exists as a triplet due to F-F coupling (J$_{FF}$=7.2 Hz). Strong downfield shifts of the ^{19}F-NMR signals of NCNFS(O)N(CH$_3$)$_2$ (147) and the diethylamino derivative (148) are not expected because of the electron-donating effect of the -NR$_2$ group (R=CH$_3$, C$_2$H$_5$). Proton-fluorine coupling has been observed for both compounds. Fluorine atoms bonded to tetravalent sulfur in (CF$_3$)$_2$CFSF$_2$-N(CH$_3$)$_2$ (151) and (CF$_3$)$_2$CFSF$_2$N(C$_2$H$_5$)$_2$ (152) show chemical shifts $\delta_{S\text{-}F}$= -9.8 and -8.2 ppm. This result is found to be a certain contrast to the shifts of compounds (147) and (148) but it is similar to the shielding behavior of the fluorine of compounds (37-40).

TABLE 10

Data of Compounds Containing Hexavalent Sulfur[a]

No.	Compound	δ_{S-F} (ppm)	ν_{S-F} (cm^{-1})	δ_X (ppm)	δ_{C-F} (ppm)	J_{SF-X} (cps)	$J_{CF-SF, X}$ (cps)	References
134	NSF_3	-69.3(3)	775 ν_2 811 ν_4	-	-	27	-	[3,28]
134a	NSF_3	-66.8(3)	-	-	-	26	-	[7,29]
136	$NSF_2-N(C_2H_5)_2$	-78.95(1)	811 ν_{as} 755 ν_s	-	-	-	-	[45,29]
136a	$NSF_2-N(C_2H_5)_2$	-79.1(1)	-	-	-	-	-	[7]
137	$NSF_2-OC_6H_5$	-75.0(1)	780 725	-	-	-	-	[29]
138	$[(CH_3)_3SiN]_2SF_2$	-77.4(1)	cplx.	-0.15(1)	-	-	-	[49]
139	$(CF_3N)_2SF_2$	-57.2(7)	883,847	-	48.2(3)	-	8	[18]
140	$(C_2F_5N)_2SF_2$	-63.1(5)	861,833	-	86.5(3, CF_2) ? (1, CF_3)	-	8	[18]
141	$(FO_2SN)_2SF_2$	-45.3(3)	780-900	-62.0(3)	-	6.9	-	[44]
142	$CF_3SF_4NF_2$	-20.5(8)	?	-70.7(3)	65.1(5)	20(SF/CF/NF)	22	[10]
143	F_B F_A S F_C F_A NF_2(Y) OCF_3(X)	-47.2(A,cplx.) -44.2(B,cplx.) -68.2(C,cplx.)	cplx.	-76.2(Y, cplx.)	55.8 (cplx.)	see original	see original	[85]

[a] No. 134-143; no SF_5 groups; numbers or symbols in parentheses: multiplicity or shifting or coupling groups; cplx. = complex; question marks = data not given or doubtful.

TABLE 11

Data of Miscellaneous Compounds

No.	Compound	δ_{S-F} (ppm)	ν_{S-F} (cm^{-1})	δ_{N-F} (ppm)	$\delta_{F,H}$ (ppm)	$J_{H,F-F}$ (Hz)	References
144	CF_3SNF_2	–	–	-103.0(1)	111.8(3)	7.2	[59]
145	$H_2NSO_2NF_2$	–	–	-37.7(1)	not det.	–	[89]
146	$H_2NSO_2NF_2 \cdot (C_6H_5)_3PO$	–	–	-38.7(1)	not det.	–	[89]
147	$NC-NFS(O)N(CH_3)_2$	-49.8(7)	778(s)	–	-3.86(2)	3.2	[86]
148	$NC-NFS(O)N(C_2H_5)_2$	-61.3(5)	751(s)	–	-3.78(8, CH$_2$) -1.41(3, CH$_3$)	2.7(CH$_2$/F)	[86]
149	$N(CH_3)_2$	-76.82(F$_A$) -66.47(F$_E$) (8)	821(vs) 709(vs)	–	-2.84(6, F$_E$, F$_A$) -1.36(3, CH$_3$)	4.2(H/F$_A$) 2.2(H/F$_E$)	[86, 90]
150	$N(C_2H_5)_2$	-74.76(F$_A$) -70.84(F$_E$) (8)	815(s) 710(vs)	–	-3.54(4, CH$_2$)	not det.	[86]

TABLE 11 (Continued)

Data of Miscellaneous Compounds

No.	Compound	δS-F (ppm)	νS-F (cm^{-1})	δN-F (ppm)	δF, H (ppm)	$J_{H, F-F}$ (Hz)	References
151	$(CF_3)_2CFSF_2N(CH_3)_2$	$-9.8(?)$	632(s) ν_s 610(s) ν_{as}	–	156.7 (?CF) 72.7 (?CF$_3$) -2.88 (?CH$_3$)	20.3(CF/SF$_2$) 12.3(CF$_3$/SF$_2$) 3.6(CH$_3$/SF$_2$)	[91]
152	$(CF_3)_2CFSF_2N(C_2H_5)_2$	$-8.2(?)$	629(vs) ν_s 605(vs) ν_{as}	–	157.2 (?CF) 72.3 (?CF$_3$) -3.61 (?CH$_2$) -1.22 (?CH$_3$)	19.5(CF/SF$_2$) 11.0(CF$_3$/SF$_2$) 2.0(CH$_2$/SF$_2$)	[91]

a No. 144–152; numbers or symbols in parentheses: multiplicity and/or shifted and/or coupling groups; 149: t = -30°C; 150: t = -40°C; 151: t = 30°C, (CF/CF$_3$) J_{FF} = 5.0 Hz, (CF'/CF$_3$) J_{FF} = 5.0 Hz, (CF/CH$_3$) J_{FH} = 0.9 Hz; 152: t = 30°C, (CF/CF$_3$) J_{FF} = 4.7 Hz, (CH$_2$/CH$_3$) J_{HH} = 7.0 Hz.

A more complicated picture appears on recording ^{19}F-NMR spectra of compounds (149) and (150). A typical A_2B spectrum results because of the magnetically nonequivalent fluorine atoms F_a (2) and F_e (1). Nine lines are expected, but only eight have been observed, the last line being too weak to be observed. A computation using programs LAOCON I and LAOCON II was carried out to calculate positions of the lines and good agreement [86] could be shown between experimental and calculated values.

B. Characteristic Frequencies

The molecules listed in Table 10 do have absorption frequencies in the usual range at which S-F vibrations do occur. Every band in the ir spectra is of strong to very strong intensity. S-F vibrations do absorb at higher wavenumbers in compounds containing sulfur of oxidation number six compared to molecules with sulfur of oxidation number four. In some cases symmetric and asymmetric vibrations were assigned to bands of the S-F region.

REFERENCES

[1]   O. Glemser, Endeavour, 28, 86 (1969).

[2]   O. Glemser and H. Richert, Z. Anorg. Allgem. Chem., 307, 313 (1961).

[3]   H. Richert and O. Glemser, Z. Anorg. Allgem. Chem., 307, 328 (1961).

[4]   H. Schröder and O. Glemser, Z. Anorg. Allgem. Chem., 298, 78 (1959).

[5]   O. Glemser, H. Schröder, and E. Wyszomirski, Z. Anorg. Allgem. Chem., 298, 72 (1959).

[6]   O. Glemser, R. Mews, and H. W. Roesky, Chem. Commun., 1969, 914.

[7]   H.-G. Horn, unpublished results.

[8]   J. K. Ruff, Inorg. Chem., 5, 1787 (1966).

[9]   B. Cohen, T. R. Hooper and R. D. Peacock, Chem. Commun., 1966, 32.

[10] A. L. Logothethis, G. N. Sausen, and R. J. Shozda, Inorg. Chem., 2, 173 (1963).

[11] A. F. Clifford and J. W. Thompson, Inorg. Chem., 5, 1424 (1966).

[12] A. F. Clifford and G. R. Zeilenga, Inorg. Chem., 8, 1789 (1969).

[13] A. F. Clifford and G. R. Zeilenga, Inorg. Chem., 8, 979 (1969).

[14] A. F. Clifford and R. G. Goel, Inorg. Chem., 8, 2004 (1969).

[15] B. Krebs, E. Meyer-Hussein, O. Glemser, and R. Mews, Chem. Commun., 1960, 1578.

[16] O. Glemser, R. Mews and H. W. Roesky, Chem. Ber., 102, 1523 (1969).

[17] O. Glemser, Prep. Inorg. Reactions, 1, 227 (1964).

[18] M. Lustig and J. K. Ruff, Inorg. Chem., 4, 1444 (1965).

[19] W. C. Smith, C. W. Tullock, R. D. Smith, and V. A. Engelhardt, J. Am. Chem. Soc., 82, 551 (1960).

[20] W. Verbeek and W. Sundermeyer, Angew. Chem., 81, 331 (1969).

[21] W. Sundermeyer, Angew. Chem., 79, 98 (1967).

[22] W. Sundermeyer and J. Stenzel, provate communication.

[23] B. Cohen and A. G. McDiarmid, J. Chem. Soc., A, 1966, 1780.

[24] G. C. Demitras and A. G. McDiarmid, Inorg. Chem., 6, 1903 (1967).

[25] D. Schläfer and M. Becke-Goehring, Z. Anorg. Allgem. Chem., 362, 1 (1968).

[26] G. Simon, Dissertation, Saarbrücken, 1964.

[27] O. Glemser, H. W. Roesky and P. R. Heinze, Angew. Chem., 79, 153 (1967).

[28] O. Glemser, H. Meyer, and A. Haas, Chem. Ber., 97, 1704 (1964).

[29]   O. Glemser and W. Koch, Z. Naturforsch., 23b, 745 (1968).

[30]   O. Glemser and S. P. v. Halasz, Z. Naturforsch., 23b, 743 (1968).

[31]   O. Glemser, H. W. Roesky, and P. R. Heinze, Angew. Chem. (Intern. Ed.) 6, 710 (1967).

[32]   A. F. Clifford and C. S. Kobayashi, Inorg. Chem., 4, 571 (1965).

[33]   J. K. Ruff, Inorg. Chem., 6, 2108 (1967).

[34]   M. Lustig, Inorg. Chem., 8, 443 (1969).

[35]   M. Lustig, C. L., C. L. Bumgardner, F. A. Johnson, and J. K. Ruff, Inorg. Chem., 3, 1165 (1964).

[36]   H. W. Roesky, Angew. Chem., 80, 44 (1968).

[37]   H. W. Roesky and A. Hoff, Chem. Ber., 101, 162 (1968).

[38]   H. W. Roesky, Angew. Chem., 79, 724 (1967).

[39]   A. Haas and P. Schott, Chem. Ber., 101, 3407 (1968).

[40]   A. F. Clifford and L. C. Duncan, Inorg. Chem., 5, 692 (1966).

[41]   G. H. Cady, D. F. Eggers and B. Tittle, Proc. Chem. Soc., 1963, 65.

[42]   C. Merrill, S. Williamson, G.H. Cady, and D. Eggers, Inorg. Chem., 1, 215 (1962).

[43]   D. Schläfer, Dissertation, Heidelberg, 1966.

[44]   H. W. Roesky and D. P. Babb, Angew. Chem., 81, 494 (1969).

[45]   O. Glemser, H. Meyer, and A. Haas, Chem. Ber., 98, 2049 (1965).

[46]   O. Glemser and S. P. v. Halasz, Chem. Ber., 102, 3333 (1969).

[47]   E. Fluck, Die Kernmagnetische Resonanz Und Ihre Anwendung In Der Anorganischen Chemie, Springer, Berlin, 1963.

[48]   M. Lustig, private communication.

[49]   O. Glemser and J. Wegener, Angew. Chemie, 82, 324 (1970).

[50]   B. Cohen and A. G. McDiarmid, Angew. Chemie, 75, 207 (1963).

[51]   R. Mews and O. Glemser, Inorg. Nucl. Chem. Letters, 6, 35 (1970).

[52]   J. E. Griffiths and D. F. Sturman, Spectrochimica Acta, 25A, 1355 (1969).

[53]   O. Glemser and S. P. v. Halasz, Inorg. Nucl. Chem. Letters, 5, 393 (1969).

[54] O. Glemser and S. P. v. Halasz, Inorg. Nucl. Chem. Letters, 4, 191 (1968).

[55] U. Biermann and O. Glemser, Chem. Ber., 100, 3795 (1967).

[56] T. Moeller and A. Ouchi, J. Inorg. Nucl. Chem., 28, 2147 (1966).

[57] T. H. Chang, T. Moeller, and C. W. Allen, J. Inorg. Nucl. Chem., 32, 1043 (1970).

[58] L. C. Duncan, Inorg. Chem., 9, 987 (1970).

[59] E. C. Stump and C. D. Padgett, Inorg. Chem., 3, 610 (1964).

[60] H. W. Roesky, H. H. Giere, and D. P. Babb, Inorg. Chem., 9, 1076 (1970).

[61] C. W. Tullock, D. D. Coffman, and E. L. Muetterties, J. Am. Chem. Soc., 86, 357 (1964).

[62] E. C. Stump, C. D. Padgett and W. S. Brey, Inorg. Chem., 2, 648 (1963).

[63] R. D. Dresdner, J. S. Johar, J. Merritt and C. S. Patterson, Inorg. Chem., 4, 678 (1965).

[64] J. S. Johar and R. D. Dresdner, Inorg. Chem., 7, 683 (1968).

[65] R. E. Noftle and J. M. Shreeve, Inorg. Chem., 7, 687 (1968).

[66] O. Glemser and U. Biermann, Inorg. Nucl. Chem. Letters, 3, 223 (1967).

[67] F. Seel and G. Simon, Z. Naturforsch. 19b, 354 (1964).

[68] G. Demitras, R. A. Kent, and A. G. McDiarmid, Chem. Industry, 1964, 1712.

[69] H. W. Roesky, G. Holtschneider and H. H. Giere, Z. Naturforsch., 25b, 252 (1969).

[70] H. W. Roesky and S. Tutkunkardes, Z. Anorg. Allgem., 374, 147 (1970).

[71] H. W. Roesky and H. H. Giere, Angew. Chem., 82, 255 (1970).

[72] H. Siebert, Anwendungen Der Schwingungsspektroskopie In Der Anorganischen Chemie, Springer, Berlin, 1966.

[73] O. Glemser, A. Müller, D. Böhler, and B. Krebs, Z. Anorg. Allgem. Chem., 357, 185 (1968).

[74] A. Müller, O. Glemser, and B. Krebs, Z. Naturforsch., 22b, 550 (1967).

[75] H. W. Roesky and U. Biermann, Angew. Chemie, 79, 904 (1967).

[76] H. W. Roesky, Inorg. Nucl. Chem. Letters, 4, 147 (1968).

[77] H. W. Roesky and D. P. Babb, Inorg. Chem., 8, 1733 (1969).

[78] H. W. Roesky and D. P. Babb, Angew. Chemie, 81, 705 (1969).

[79] L. K. Huber and H. C. Mandell, Inorg. Chem., 4, 919 (1965).

[80] R. Appel and G. Eisenhauer, Chem. Ber., 95, 246 (1962).

[81] H. W. Roesky, Angew. Chem., 80, 626 (1968).

[82] H. W. Roesky and H. H. Giere, Chem. Ber., 102, 3707 (1969).

[83] H. W. Roesky, Z. Anorg. Allgem. Chem., 367, 151 (1969).

[84] H. Eysel, J. Mol. Structure, 5, 275 (1970).

[85] L. C. Duncan and G. H. Cady, Inorg. Chem., 3, 1045 (1964).

[86] S. P. v. Halasz and O. Glemser, Chem. Ber., 103, 594 (1970).

[87] S. P. v. Halasz and O. Glemser, Chem. Ber., 103, 553 (1970).

[88] O. Glemser, S. P. v. Halasz, and U. Biermann, Inorg. Nucl. Chem. Letters, 4, 591 (1968).

[89] R. A. Wiesboeck and J. K. Ruff, Inorg. Chem., 4, 123 (1965).

[90] O. Glemser, S. P. v. Halasz, and U. Biermann, Z. Naturforsch., 23b, 1381 (1968).

[91] R. Mews, G. G. Alange, and O. Glemser, Naturwissensch., 57, 245 (1968).

[92] K. Seppelt and W. Sundermeyer, Angew. Chemie, 82 (1970) (in the press).

[93] O. Glemser, R. Mews, and S. P. v. Halasz, Inorg. Nucl. Chem. Letters, 5, 321 (1969).

[94] G. W. Parshall, R. Cramer, and R. E. Forster, Inorg. Chem., 1, 677 (1962).

[95] R. Cramer and D. D. Coffman, J. Org. Chem., 26, 4010 (1961).

[96] W. Sundermeyer, private communication.

[97] O. Glemser, U. Biermann and A. Hoff, Z. Naturforsch., 22b, 893 (1967).

[98]   A. J. Banister, L. F. Moore and J. S. Padley, Spectrochim. Acta,
       23A, 2705 (1967).

[99]   A. J. Banister and B. Bell, J. Chem. Soc. , 1970, A, 1659.

AUTHOR INDEX

Numbers in parentheses are reference numbers and indicate that an author's work is referred to although his name is not cited in the text. Underlined numbers give the page on which the complete reference is listed.

A

B

SUBJECT INDEX

A

Alicyclic compounds, polyfluorinated, radiation chemistry,
31-32
Aromatic compounds, polyfluorinated, radiation chemistry,
33

B

Bromotrifluoromethane, radiation
chemistry, 14 ff., 22

C

Characteristic frequencies of N-S-F
compounds, 140-142
compounds containing N-S-F
bonds, 150, 154, 156-157,
159-178, 181, 182, 187
Chelating agents, fluoro-β-
diketones as, 57-117
Chemical shift of N-S-F
compounds, 137-139
Chlorotrifluoromethane, radiation
chemistry, 15 ff.
Claisen condensation, for fluoro-β-
diketones, 45-46
Compton effect, 5-6
Coupling constants of N-S-F
compounds, 139-140

D

Dibromodifluoromethane, radiation
chemistry, 14 ff.
β-Diketones, fluorine-substituted,
43-117

E

Electron capture, 33-39
Electron effects, secondary, 6-7
Excited molecules, reactions of,
7-11

F

Fluoro-β-diketones, 43 ff.
chemical reactivity, 47-49
as metal chelating agents, 57-117
miscellaneous properties, 57
physical properties, 52-56
synthesis, 45-47
synthetic applications not involving metals, 49-52
Fluoroolefins, radiation chemistry,
28 ff.

G

Gamma rays, interaction with
matter, 3-4
Gas chromatography, of fluoro-β-
diketonates, 92, 112-114

H

Halofluoromethanes, radiation
chemistry, 14-22
Hexafluorobenzene, radiation
chemistry, 33
Hexafluoroethane, radiation
chemistry, 23-27

I

Ions, reactions in radiation
chemistry, 7-11

203